T0140029

Advances in Intelligent Systems and Computing

Volume 803

Series editor

Janusz Kacprzyk, Polish Academy of Sciences, Warsaw, Poland
e-mail: kacprzyk@ibspan.waw.pl

The series "Advances in Intelligent Systems and Computing" contains publications on theory, applications, and design methods of Intelligent Systems and Intelligent Computing. Virtually all disciplines such as engineering, natural sciences, computer and information science, ICT, economics, business, e-commerce, environment, healthcare, life science are covered. The list of topics spans all the areas of modern intelligent systems and computing such as: computational intelligence, soft computing including neural networks, fuzzy systems, evolutionary computing and the fusion of these paradigms, social intelligence, ambient intelligence, computational neuroscience, artificial life, virtual worlds and society, cognitive science and systems, Perception and Vision, DNA and immune based systems, self-organizing and adaptive systems, e-Learning and teaching, human-centered and human-centric computing, recommender systems, intelligent control, robotics and mechatronics including human-machine teaming, knowledge-based paradigms, learning paradigms, machine ethics, intelligent data analysis, knowledge management, intelligent agents, intelligent decision making and support, intelligent network security, trust management, interactive entertainment, Web intelligence and multimedia.

The publications within "Advances in Intelligent Systems and Computing" are primarily proceedings of important conferences, symposia and congresses. They cover significant recent developments in the field, both of a foundational and applicable character. An important characteristic feature of the series is the short publication time and world-wide distribution. This permits a rapid and broad dissemination of research results.

Advisory Board

Chairman

Nikhil R. Pal, Indian Statistical Institute, Kolkata, India
e-mail: nikhil@isical.ac.in

Members

Rafael Bello Perez, Universidad Central "Marta Abreu" de Las Villas, Santa Clara, Cuba
e-mail: rbellop@uclv.edu.cu

Emilio S. Corchado, University of Salamanca, Salamanca, Spain
e-mail: escorchado@usal.es

Hani Hagras, University of Essex, Colchester, UK
e-mail: hani@essex.ac.uk

László T. Kóczy, Széchenyi István University, Győr, Hungary
e-mail: koczy@sze.hu

Vladik Kreinovich, University of Texas at El Paso, El Paso, USA
e-mail: vladik@utep.edu

Chin-Teng Lin, National Chiao Tung University, Hsinchu, Taiwan
e-mail: ctlin@mail.nctu.edu.tw

Jie Lu, University of Technology, Sydney, Australia
e-mail: Jie.Lu@uts.edu.au

Patricia Melin, Tijuana Institute of Technology, Tijuana, Mexico
e-mail: epmelin@hafsamx.org

Nadia Nedjah, State University of Rio de Janeiro, Rio de Janeiro, Brazil
e-mail: nadia@eng.uerj.br

Ngoc Thanh Nguyen, Wroclaw University of Technology, Wroclaw, Poland
e-mail: Ngoc-Thanh.Nguyen@pwr.edu.pl

Jun Wang, The Chinese University of Hong Kong, Shatin, Hong Kong
e-mail: jwang@mae.cuhk.edu.hk

More information about this series at http://www.springer.com/series/11156

Florentino Fdez-Riverola
Mohd Saberi Mohamad · Miguel Rocha
Juan F. De Paz · Pascual González
Editors

Practical Applications of Computational Biology and Bioinformatics, 12th International Conference

 Springer

Editors
Florentino Fdez-Riverola
Escuela Superior de Ingeniería Informática
Universidad de Vigo
Ourense, Spain

Mohd Saberi Mohamad
Faculty of Computing, Department
  of Software Engineering
Universiti Teknologi Malaysia
Johor, Malaysia

Miguel Rocha
Department de Informática
Universidade do Minho
Braga, Portugal

Juan F. De Paz
Departamento de Informática y Automática,
  Facultad de Ciencias
Universidad de Salamanca
Salamanca, Spain

Pascual González
Departamento de Sistemas Informáticos
Universidad de Castilla-La Mancha
Albacete, Albacete
Spain

ISSN 2194-5357   ISSN 2194-5365 (electronic)
Advances in Intelligent Systems and Computing
ISBN 978-3-319-98701-9   ISBN 978-3-319-98702-6 (eBook)
https://doi.org/10.1007/978-3-319-98702-6

Library of Congress Control Number: 2018954643

© Springer Nature Switzerland AG 2019
This work is subject to copyright. All rights are reserved by the Publisher, whether the whole or part of the material is concerned, specifically the rights of translation, reprinting, reuse of illustrations, recitation, broadcasting, reproduction on microfilms or in any other physical way, and transmission or information storage and retrieval, electronic adaptation, computer software, or by similar or dissimilar methodology now known or hereafter developed.
The use of general descriptive names, registered names, trademarks, service marks, etc. in this publication does not imply, even in the absence of a specific statement, that such names are exempt from the relevant protective laws and regulations and therefore free for general use.
The publisher, the authors, and the editors are safe to assume that the advice and information in this book are believed to be true and accurate at the date of publication. Neither the publisher nor the authors or the editors give a warranty, express or implied, with respect to the material contained herein or for any errors or omissions that may have been made. The publisher remains neutral with regard to jurisdictional claims in published maps and institutional affiliations.

This Springer imprint is published by the registered company Springer Nature Switzerland AG
The registered company address is: Gewerbestrasse 11, 6330 Cham, Switzerland

Preface

Next generation sequencing technologies, together with other emerging and quite diverse experimental techniques are evolving rapidly, creating numerous types of omics data. These are creating new challenges for the expanding fields of Bioinformatics and Computational Biology, which seek to analyse, process, integrate and extract meaningful knowledge from these data. This calls for new algorithms and approaches from fields such as Databases, Statistics, Data Mining, Machine Learning, Optimization, Computer Science, Machine Learning and Artificial Intelligence. Clearly, Biology is increasingly becoming a science of information, requiring tools from the computational sciences. To address these challenges, we have seen the surge of a new generation of interdisciplinary scientists with a strong background in the biological and computational sciences.

The International Conference on Practical Applications of Computational Biology & Bioinformatics (PACBB) is an annual international meeting dedicated to emerging and challenging applied research in Bioinformatics and Computational Biology. Building on the success of previous events, the 12th edition of PACBB Conference will be held on 20–22 June 2018 in the University of Castilla-La Mancha, Toledo (Spain). In this occasion, special issues will be published by the Interdisciplinary Sciences-Computational Life Sciences, Journal of Integrative Bioinformatics, Neurocomputing, Knowledge and Information Systems: An International Journal covering extended versions of selected articles.

This volume gathers the accepted contributions for the 12th edition of the PACBB Conference after being reviewed by different reviewers, from an international committee from 13 countries. PACBB'18 technical program includes 26 papers spanning many different sub-fields in Bioinformatics and Computational Biology.

Therefore, this event will strongly promote the interaction of researchers from diverse fields and distinct international research groups. The scientific content will be challenging and will promote the improvement of the valuable work that is being carried out by the participants. In addition, it will promote the education of young scientists, in a post-graduate level, in an interdisciplinary field.

We would like to thank all the contributing authors and sponsors, as well as the members of the Program Committee and the Organizing Committee for their hard and highly valuable work and support. Their effort has helped to contribute to the success of the PACBB'18 event. PACBB'18 wouldn't exist without your assistance.

Mohd Saberi Mohamad
PACBB'18 Programme Co-chairs
Miguel P. Rocha
PACBB'18 Programme Co-chairs
Juan F. De Paz
PACBB'18 Programme Co-chairs
Florentino Fdez-Riverola
PACBB'18 Organizing Co-chairs
Pascual González
PACBB'18 Organizing Co-chairs

Organization

General Co-chairs

Mohd Saberi Mohamad   Universiti Malaysia Kelantan, Malaysia
Miguel Rocha   University of Minho, Portugal
Juan F. De Paz   University of Salamanca, Spain
Florentino Fdez-Riverola   University of Vigo, Spain
Pascual González   University of Castilla-La Mancha, Spain

Program Committee

Alberto López   University of Salamanca, Spain
Alejandro F. Villaverde   IIM-CSIC, Spain
Alexandre Perera Lluna   Universitat Politècnica de Catalunya, Spain
Alfonso Rodriguez-Paton   Universidad Politecnica de Madrid, Spain
Alfredo Vellido Alcacena   UPC, Spain
Alicia Troncoso   University Pablo de Olavide, Spain
Álvaro Lozano   University of Salamanca, Spain
Amin Shoukry   Egypt-Japan University of Science and Technology, Egypt
Amparo Alonso   University of A Coruña, Spain
Ana Cristina Braga   University of Minho, Portugal
Ana Margarida Sousa   University of Minho, Portugal
Anália Lourenço   University of Vigo, Spain
André Sales   University of Salamanca, Spain
Armando Pinho   Universty of Aveiro, Portugal
Boris Brimkov   Rice University, USA
Carlos A. C. Bastos   University of Aveiro, Portugal
Carole Bernon   IRIT/UPS, France

Carolyn Talcott	Stanford University, USA
Daniel Glez-Peña	University of Vigo, Spain
Daniel Hernández	University of Salamanca, Spain
David Hoksza	Charles University in Prague, Czech Republic
David Rodríguez Penas	IIM-CSIC, Spain
Diego M. Jiménez	University of Salamanca, Spain
Eduardo Valente	IPCB, Spain
Eva Lorenzo Iglesias	University of Vigo, Spain
Fernanda Brito Correia	University of Aveiro, Portugal
Fernando Diaz-Gómez	University of Valladolid, Spain
Filipe Liu	University of Minho, Portugal
Francisco Couto	University of Lisboa, Portugal
Gabriel Villarrubia	University of Salamanca, Spain
Gael Pérez Rodríguez	University of Vigo, Spain
Giovani Librelotto	Federal University of Santa Maria, Brasil
Gustavo Isaza	University of Caldas, Colombia
Gustavo Santos-García	University of Salamanca, Spain
Hugo López-Fernández	University of Vigo, Spain
Isabel C. Rocha	University of Minho, Portugal
Javier Bajo	Technical University of Madrid, Spain
Javier Caridad	University of Salamanca, Spain
Javier De Las Rivas	CSIC, Spain
Javier Pérez	University of Salamanca, Spain
João Ferreira	University of Lisboa, Portugal
Joel P. Arrais	DEI/CISUC University of Coimbra, Portugal
Jorge Vieira	IBMC, Porto, Portugal
José Antonio Castellanos Garzón	University of Salamanca, Spain
José Luis Oliveira	Universty of Aveiro, Portugal
Josep Gómez	Universitat Rovira i Virgili, Spain
Juan Ramos	University of Salamanca, Spain
Julio R. Banga	IIM-CSIC, Spain
Loris Nanni	University of Bologna, Italy
Lourdes Borrajo Diz	University of Vigo, Spain
Lucía Martín	University of Salamanca, Spain
Luis F. Castillo	University of Caldas, Colombia
Luis M. Rocha	Indiana University, USA
Mª Araceli Sanchís de Miguel	University of Carlos III, Spain
Manuel Álvarez Díaz	University of A Coruña, Spain
Marcelo Maraschin	Federal University of Santa Catarina, Florianopolis, Brazil
Marcos Martinez-Romero	Stanford University, UK
María Navarro	University of Salamanca, Spain
Maria Olivia Pereira	IBB - CEB Centre of Biological Engineering, Portugal

Martin Krallinger	CNIO, Spain
Martín Pérez-Pérez	University of Vigo, Spain
Masoud Daneshtalab	University of Turku, Finland
Miguel Reboiro	University of Vigo, Spain
Mohd Firdaus Raih	National University of Malaysia, Malaysia
Narmer Galeano	Cenicafé, Colombia
Nuno F. Azevedo	University of Porto, Portugal
Nuno Fonseca	CRACS/INESC, Porto, Portugal
Oscar Dias	CEB/IBB, Universidade do Minho, Portugal
Patricia González	University of A Coruña, Computer Architecture Group (GAC), Spain
Paula Jorge	IBB - CEB Centre of Biological Engineering, Portugal
Pedro G. Ferreira	Ipatimup - Institute of Molecular Pathology and Immunology of the University of Porto, Portugal
Pierpaolo Vittorini	University of L'Aquila, Italy
Ramón Doallo	University of A Coruña, Spain
René Alquezar Mancho	UPC, Spain
Rita Ascenso	Polytecnic Institute of Leiria, Portugal
Rita Margarida Teixeira Ascenso	ESTG – IPL, Portugal
Rosalía Laza	University of Vigo, Spain
Rui Camacho	Universty of Porto, Portugal
Ruben Martín	University of Salamanca, Spain
Sara C. Madeira	IST/INESC ID, Lisbon, Portugal
Sérgio Deusdado	Polytecnic Institute of Bragança, Portugal
Sergio Matos	DETI/IEETA, Portugal
Thierry Lecroq	Univeristy of Rouen, France
Turki Turki	King Abdulaziz University, Saudi Arabia
Valentin Brimkov	SUNY Buffalo State College, USA
Vera Afreixo	University of Aveiro, Portugal
Yeray Mezquita	University of Salamanca, Spain
Yingbo Cui	National University of Defense Technology, China

PACBB 2018 Sponsors

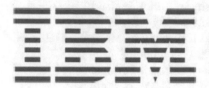

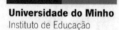

Contents

Contents

A Hybrid of Simple Constrained Artificial Bee Colony Algorithm and Flux Balance Analysis for Enhancing Lactate and Succinate in *Escherichia Coli*

Mei Kie Hon[1], Mohd Saberi Mohamad[2,3(✉)],
Abdul Hakim Mohamed Salleh[1], Yee Wen Choon[1],
Muhammad Akmal Remli[1], Mohd Arfian Ismail[4], Sigeru Omatu[5],
and Juan Manuel Corchado[6]

[1] Artificial Intelligence and Bioinformatics Research Group,
Faculty of Computing, Universiti Teknologi Malaysia,
81310 Skudai, Johor, Malaysia
{mkhon4, ywchoon2}@live.utm.my,
abdhakim.utm@gmail.com, akmalmuhd@gmail.com
[2] Institute for Artificial Intelligence and Big Data, Universiti Malaysia Kelantan,
City Campus, Pengkalan Chepa, 16100 Kota Bharu, Kelantan, Malaysia
[3] Faculty of Bioengineering and Technology, Universiti Malaysia Kelantan,
Jeli Campus, Lock Bag 100, 17600 Jeli, Kelantan, Malaysia
saberi@umk.edu.my
[4] Soft Computing and Intelligent System Research Group, Faculty of Computer
Systems and Software Engineering, Universiti Malaysia Pahang,
26300 Kuantan, Pahang, Malaysia
arfian@ump.edu.my
[5] Department of Electronics, Information and Communication Engineering,
Osaka Institute of Technology, Osaka 535-8585, Japan
omatu@rsh.oit.ac.jp
[6] Biomedical Research Institute of Salamanca/BISITE Research Group,
University of Salamanca, Salamanca, Spain
corchado@usal.es

Abstract. In the past decades, metabolic engineering has received great attention from different sectors of science due to its important role in enhancing the over expression of the target phenotype by manipulating the metabolic pathway. The advent of metabolic engineering has further laid the foundation for computational biology, leading to the development of computational approaches for suggesting genetic manipulation. Previously, conventional methods have been used to enhance the production of lactate and succinate in *E. coli*. However, these products are always far below their theoretical maxima. In this research, a hybrid algorithm is developed to seek optimal solutions in order to increase the overproduction of lactate and succinate by gene knockout in *E. coli*. The hybrid algorithm employed the Simple Constrained Artificial Bee Colony (SCABC) algorithm, using swarm intelligence as an optimization algorithm to optimize the objective function, where lactate and succinate productions are maximized by simulating gene knockout in *E. coli*. In addition, Flux Balance

© Springer Nature Switzerland AG 2019
F. Fdez-Riverola et al. (Eds.): PACBB 2018, AISC 803, pp. 1–8, 2019.
https://doi.org/10.1007/978-3-319-98702-6_1

Analysis (FBA) is used as a fitness function in the SCABC algorithm to assess the growth rate of *E. coli* and the productivity of lactate and succinate. As a result of the research, the gene knockout list which induced the highest production of lactate and succinate is obtained.

Keywords: Gene knockout strategies · Escherichia coli · Lactate
Succinate · Simple Constrained Artificial Bee Colony · Flux Balance Analysis
Computational intelligence

1   Introduction

In recent decades, many studies have been carried out on *E. coli* to increase the overproduction of lactate and succinate [1–3]. Gene knockout technology is a technique used to enhance the microbial phenotype by inactivating the gene of an organism, where the function of the gene is repressed by altering the gene [4]. The term knockout refers to creating a new mutant by "knocking out" a gene, which is essentially the opposite of a gene knock in. However, using the gene knockout technique to identify the desired gene is not straightforward as it involves a large number of interacting reactions. Instead of genetically engineering the strain, the computational method is introduced to identify the genes for manipulation, which can facilitate the overproduction of biochemical compounds as it can greatly save both time and resources [5–7].

OptKnock is the first rational modelling framework proposed, with the intention of suggesting gene knockout strategies for enhancement of the overproduction of metabolite while recognizing that the metabolic flux distributions are regulated by internal cellular objectives [8]. OptKnock was used to search for a set of gene deletions by optimizing the flux towards a targeted metabolite, while the internal flux distribution is functioned for optimizing the growth rate of the cell. In the study by Burgard [8], two objectives were in focus, specifically, the maximization of the biochemical products and the Minimization of Metabolic Adjustment (MoMA). Therefore, the OptKnock algorithm is known as a bi-level optimization framework and is implemented by using mixed integer linear programming (MILP).

In recent years, many researchers have applied a swarm intelligence algorithm as an optimization tool to suggest the gene knockout strategy. Swarm intelligence is defined as the emergent collective intelligence of groups of simple agents by imitating the group foraging of social insects, their division of labour or collective sorting and clustering [9]. Numerous algorithms have been developed, such as the Particle Swarm Optimization algorithm, Ant Colony Optimization algorithm, and Bee algorithm. The good performance of swarm intelligence to solve the local minima problem and to search for the global optimum has encouraged many researchers to apply this in gene knockout to maximize the production rate of the targeted biochemical compound and the growth rate of the microorganism [10].

In this paper, a hybrid of the Simple Constrained Artificial Bee Colony (SCABC) and Flux Balance Analysis (FBA) [11], known as SCABCFBA, is proposed to identify the optimal gene knockout strategy for enhancing the production of lactate and succinate. SCABC is an extension of the Artificial Bee Colony algorithm (ABC) and was proposed by Brajevic [12] to solve the constrained optimization problem, whereby collective intelligence imitates a population of honey bees to solve optimization problems. The improvement in SCABC is in increasing the exploration and exploitation of the solution, such that it can quickly search for the global optimum. FBA is used as a fitness function in the hybrid algorithm to assess the growth rate of *E. coli* and the productivity of lactate and succinate.

2 Materials and Methods

2.1 Dataset

In this research, the *E. coli* core model dataset is implemented to identify the gene knockout. The model is originally existed in the COBRA Toolbox and is installed in MATLAB. The dataset is named *'ecoli_core_model.mat'* and is used as direct input in the research. This model has been used in previous works, such as the studies by Burgard [8] and Rocha [13]. This model comprises a total of 26 data fields. Since the goal of the research is mainly to focus on the metabolite production for the gene knockout in *E. coli*, only 4 fields are employed in the system, which are *rxns, lb, ub and rxnNames*. The *E. coli* core model contains 95 reactions and it was inputted into the hybrid algorithm, SCABCFBA, to identify the best knockout strategies. SCABCFBA search the combination of reactions that can be deleted in order to identify high yield of biochemical productions.

2.2 A Hybrid of Simple Constrained Artificial Bee Colony Algorithm and Flux Balance Analysis

The proposed algorithm is a hybrid of SCABC and FBA, where the SCABC algorithm plays an important role in the optimization search for the best gene knockout, while FBA is performed as the fitness function in SCABC to forecast the growth rate of *E. coli* and the production rate of lactate and succinate in *E. coli*. SCABC is an extension of the ABC algorithm, so it has a similar backbone to traditional ABC. Figure 1 shows the pseudo code of the hybrid algorithm respectively. The algorithm is divided into five phases: (i) Initialize the Population, (ii) Employed Phase, (iii) Onlooker Phase, (iv) Memorize the Best, and (v) Scout Phase.

Initialization Phase
In the first phase, the algorithm starts by initializing a population with the matrix $i \times j$, where i denotes the number of reactions in the *E. coli* core model and j denotes the number of possible solutions in the population. The matrix is initialized with all the value 0s, whereby it indicates the *E. coli* model consisting of all the reactions at the beginning. Then the value 1s is randomly distributed in the matrix, which is essential as it represents that the corresponding reactions are removed.

After the population has been initialized, the lists of reactions are used and inputted into the FBA to get the fitness value of each possible solution. The growth rate of

1. Initialize the population solutions x_{ij} for the first run. For every other run, if exists,
   x_{i1}, is the best solution from previous run and $x_{ij}, j = 2,...,100$ are randomly
   produced solutions.
2. Evaluate fitness value of the population with FBA
3. cycle = 1
4. **repeat**
5. Produce new solutions *for* the employed bees evaluate them with FBA
6. Apply selection process based on Tournament Selection
7. Calculate the probability values Pk for the solutions x_{ij} where k=1,2,..., j.
8. Produce the new solutions for the onlookers from the solutions x_{ij} selected depending
   on Pk and evaluate their fitness value with FBA
9. Apply selection process based on Tournament Selection
10. Determine the abandoned feasible solution for the scout, if exists, and replace it with
    a new randomly produced solution x_{ij}
11. Every infeasible solution replace with randomly produced solution x_{ij}
12. Memorize the best solution achieved so far
13. cycle = cycle + 1
14. **until** cycle = MaximumCycleNumber
15. Return best solution

Fig. 1. Pseudo code of SCABCFBA algorithm

E. coli and the minimum production of the targeted metabolites for the corresponding population are evaluated by FBA.

Employed Phase

In the employed phase, a new population is randomly created with the same size as the original population. The fitness of the new population is evaluated by using FBA. Based on the fitness obtained from the original and new populations, tournament selection is employed as the selection operator between the original and new solution populations. In this study, two constraints are employed, which are the growth rate of the cell and the minimum production of the targeted metabolite. The growth rate is used to determine whether the cell survives after gene deletion, where a negative value of growth rate indicates that the cell cannot survive while a positive value indicates that the cell is able to survive and reproduce after the gene is removed. It is constrained such that the feasible solution must be more than 0.1 to make sure that the cell survives after gene knockout. To prevent very small values of the minimum production of metabolite, the constraint of minimum production is defined as being more than $-1e^{-3}$. Based on the constraints defined, tournament selection is employed to seek the optimal solution [12].

Onlooker Phase

In the onlooker phase, a new population is formed by selecting a possible solution with the probability from the populations generated in the employed phase. The value of probability P is calculated by using the formula:

$$P_{k=\frac{fit_k}{\sum_{k=1}^{j} fit}} \tag{1}$$

where j denotes the number of populations. The probability of the population solution is weighted by the possibility of a solution to be selected. If the probability of a solution is high, the solution will have a high possibility as the input in this phase. In consequence, a new population solution is generated via the tournament selection, where the best solution is selected between the two populations, and the solution having the higher fitness will replace the old solution in the population. Similar to the employed phase, the control parameter is used to limit the abandonment of the population. The trial value will be increased by 1 if the fitness of the population is not improved, whereas a replaced population will remain the trial value to 0.

Scout Phase

When there is a solution that cannot be further improved through the predetermined control parameter, the solution is considered exhausted and it will be abandoned. Once the control parameter which is the number of trials exceeds the limit, defined in this study as 50 cycles, the scout phase is invoked by randomly generating another new solution to replace all the abandoned solutions. In addition, the scout phase also checks all the solutions to determine if any solution is infeasible. If the solution is not feasible, a new solution is randomly created to replace the infeasible solution. The new solution population is then calculated by FBA to get the new fitness and evaluated to select the best solution amongst two populations.

Memorize the Best Phase

In the last phase, the reaction knockout list is returned as a result. All the phases in the system are repeated again and go through the population initialization phase, employed phase, onlooker phase, memorize the best phase and scout phase if any abandoned solution exists. The system is repeated until it satisfies the termination criterion, where it reaches the maximum number of cycles. For the other runs, the best solution from the previous run will be initialized as the first new solution of the population in the initialization phase.

At the end, the best knockout reaction list in the population which has the best fitness value is returned as the result, denoting the best gene knockout strategy in terms of local and global search in the SCABC algorithm.

3 Results and Discussion

To show the effectiveness of our proposed method, the obtained results were validated by comparing them with previous studies. Table 1 displays a comparison between the results of OptKnock and SCABCFBA, where the knockout lists that achieved the

Table 1. Comparative experimental results of OptKnock and SCABCFBA for lactate and succinate.

Methods	Lactate		Succinate	
	Gene Knockouts	Production (mmol gDW⁻¹ hr⁻¹)	Gene Knockouts	Production (mmol gDW⁻¹ hr⁻¹)
OptKnock	*ackA, pfkA, pfkB*	18.00	*pflB, ldhA*	10.7
SCABCFBA	*gdhA, adhE, pntAB*	18.1761	*pgi,sfcA,icd*	12.1932

Shaded cells represent the best results

highest production of lactate and succinate from each of the two methods are obtained and compared. The results for OptKnock are taken from the paper by Burgard [8].

As shown in Table 1, the highest lactate production achieved by OptKnock is 18.00 (mmol gDW^{-1} hr^{-1}), which is lower than the result computed by SCABCFBA, which expressed 18.1761(mmol gDW^{-1} hr^{-1}) for lactate production. As for the succinate production, OptKnock expressed 10.7 (mmol gDW^{-1} hr^{-1}), whereas SCABCFBA showed a higher succinate production compared with OptKnock, achieving 12.19332 (mmol gDW^{-1} hr^{-1}). The obtained results prove that the hybrid algorithm, SCABCFBA, has excellent performance compared with OptKnock for identifying gene knockout strategies for metabolite overproductions.

The obtained results were also compared with wet laboratory journals based on the metabolites production. Table 2 shows the productivity of lactate from the engineered strains identified in Yang [14]. From Table 2, the highest production of lactate is YBS125, where it has expressed 8.29 (mmol gDW^{-1} hr^{-1}). The lactate production of YBS125 is higher than YBS142 and YBS121 which yield 0.06 (mmol gDW^{-1} hr^{-1}) and 5.13 (mmol gDW^{-1} hr^{-1}) of lactate production respectively. Compared with the result obtained from SCABCFBA, the highest production rate of lactate was achieved by mutants B and C, both of which expressed 18.1761 (mmol gDW^{-1} hr^{-1}) for lactate production, which is greater than the results acquired from the wet laboratory journal.

Table 2. Lactate yields from the engineered strain in *E. coli* [14].

Strains	Relevant deletions	Lactate (mmol gDW^{-1} hr^{-1})
YBS142	*nuo*	0.06
YBS121	*ackA, pta*	5.13
YBS125	*ackA, pta, nuo*	8.29

Table 3 shows the productivity of succinate from the engineered strains in *E. coli*. The experimental data for the engineered strains is taken from the paper in [15]. The succinate production obtained from the previous work [15] is the average production of succinate by knocking out a single gene using *E. coli* BW25113. From Table 3, the highest production of succinate is achieved by deactivating the *pykF* gene, which has expressed 1.54 (mmol gDW^{-1} hr^{-1}). Compared with the result from this research, the highest production rate of succinate was achieved by mutant F, which expressed 12.1932 (mmol gDW^{-1} hr^{-1}) with the deletion of the genes *pgi, icd and sfcA*. Based on the comparative study using the wet laboratory journal and OptKnock, SCABCFBA showed its reliability whereby the results achieved by SCABCFBA are higher than the wet laboratory results and OptKnock.

Table 3. Succinate yields from the engineered strain in *E. coli* [15].

Relevant deletions	Succinate (mmol gDW^{-1} hr^{-1})
pta	0.77
ppc	0.00
adhE	0.07
pykF	1.54

4 Conclusion

In this research, the hybrid algorithm known as SCABCFBA is proposed to identify the optimal solution for suggesting gene knockout in order to improve the overproduction of lactate and succinate in *E. coli*. From the results, SCABCFBA has achieved a better performance than OptKnock, suggesting that it is a feasible solution for gene knockouts. Our proposed method is based on the SCABC algorithm, which is capable of performing local and global searching simultaneously and has multivariable function optimization to solve constrained problems. The improvement made by the ABC algorithm increases the exploration and exploitation of the solution, with such improvement enhancing the convergence speed to enable a quick search towards the global optimum.

In the future, we suggest that Minimization of Metabolic Adjustment (MoMA) should be implemented in SCABC as a fitness function. MoMA is used to predict the metabolic steady state after gene knockouts. The main objective of MoMA is to minimize the Euclidean Distance of flux distribution between the metabolic states of the mutant and wild-type, and therefore a better algorithm and result may be found by implementing MoMA into SCABC.

Acknowledgement. We would like to thank Malaysian Ministry of Higher Education and Universiti Teknologi Malaysia for supporting this research by the Fundamental Research Grant Schemes (grant number: R.J130000.7828.4F886 and R.J130000.7828.4F720). We would also like to thank Universiti Malaysia Pahang for sponsoring this research via the RDU Grant (Grant Number: RDU180307).

References

1. Vemuri, G.N., Eiteman, M.A., Altman, E.: Effects of growth mode and pyruvate carboxylase on succinic acid production by metabolically engineered strains of Escherichia coli. Appl. Environ. Microbiol. **68**, 1715–1727 (2002)
2. Jantama, K., Zhang, X., Moore, J.C., Shanmugam, K.T., Svoronos, S.A., Ingram, L.O.: Eliminating side products and increasing succinate yields in engineered strains of Escherichia coli C. Biotechnol. Bioeng. **101**, 881–893 (2008)
3. Zhou, L., Zuo, Z.-R., Chen, X.-Z., Niu, D.-D., Tian, K.-M., Prior, B.A., Shen, W., Shi, G.-Y., Singh, S., Wang, Z.-X.: Evaluation of genetic manipulation strategies on D-lactate Production by Escherichia coli. Curr. Microbiol. **62**, 981–989 (2011)
4. Lee, S., Lee, D., Kim, T., Kim, B.: Metabolic engineering of Escherichia coli for enhanced production of succinic acid, based on genome comparison and in silico gene knockout simulation. Appl. Environ. Microbiol. **71**, 7880–7887 (2005)
5. Salleh, A.H.M., Mohamad, M.S., Deris, S., Omatu, S., Fdez-Riverola, F., Corchado, J.M.: Gene knockout identification for metabolite production improvement using a hybrid of genetic ant colony optimization and flux balance analysis. Biotechnol. Bioprocess Eng. **20**, 685–693 (2015)
6. Choon, Y.W., Mohamad, M.S., Deris, S., Illias, R.M., En Chai, L., Chong, C.K.: Identifying gene knockout strategy using Bees Hill Flux Balance Analysis (BHFBA) for improving the production of ethanol in bacillus subtilis. In: Advances in Biomedical Infrastructure 2013. Studies in Computational Intelligence, vol. 477, pp. 117–126. Springer, Heidelberg (2013)
7. Choon, Y.W., Mohamad, M.S., Deris, S., Illias, R.M., Chong, C.K., Chai, L.E., Omatu, S., Corchado, J.M.: Differential bees flux balance analysis with OptKnock for in silico microbial strains optimization. PLoS One **9**(7), e102744 (2014)
8. Burgard, A.P., Pharkya, P., Maranas, C.D.: OptKnock: a bilevel programming framework for identifying gene knockout strategies for microbial strain optimization. Biotechnol. Bioeng. **84**, 647–657 (2003)
9. Martino, G.D.S., Cardillo, F.A., Starita, A.: A new swarm intelligence coordination model inspired by collective prey retrieval and its application to image alignment. In: Runarsson, T. P., Beyer, H.-G., Burke, E., Merelo-Guervós, J.J., Whitley, L.D., Yao, X. (eds.) PPSN 2006. LNCS, vol. 4193, pp. 691–700. Springer, Heidelberg (2006). https://doi.org/10.1007/11844297_70
10. Patil, K., Rocha, I., Förster, J., Nielsen, J.: Evolutionary programming as a platform for in silico metabolic engineering. BMC Bioinform. **6**, 308 (2005)
11. Raman, K., Chandra, N.: Flux balance analysis of biological systems: applications and challenges. Brief Bioinform. **10**(4), 435–449 (2009)
12. Brajevic, I., Tuba, M., Subotic, M.: Performance of the improved artificial bee colony algorithm on standard engineering constrained problems. Int. J. Math. Comput. Simul. **5**, 135–143 (2011)
13. Rocha, M., Maia, P., Mendes, R., Pinto, J.P., Ferreira, E.C., Nielsen, J., Patil, K., Rocha, I.: Natural computation meta-heuristics for the in silico optimization of microbial strains. BMC Bioinform. **9**, 499 (2008)
14. Yang, Y.T., Bennett, G.N., San, K.Y.: Effect of inactivation of nuo and ackA-pta on redistribution of metabolic fluxes in Escherichia coli. Biotechnol. Bioeng. **65**, 291–297 (1999)
15. Zhu, J., Shimizu, K.: Effect of a single-gene knockout on the metabolic regulation in Escherichia coli for D-lactate production under microaerobic condition. Metab. Eng. **7**, 104–115 (2005)

Parameter Estimation of Essential Amino Acids in *Arabidopsis thaliana* Using Hybrid of Bees Algorithm and Harmony Search

Mei Yee Aw[1], Mohd Saberi Mohamad[2,3(✉)], Chuii Khim Chong[1], Safaai Deris[2,3], Muhammad Akmal Remli[1], Mohd Arfian Ismail[4], Juan Manuel Corchado[5], and Sigeru Omatu[6]

[1] Artificial Intelligence and Bioinformatics Research Group, Faculty of Computing, Universiti Teknologi Malaysia, 81310 Skudai, Johor, Malaysia
{myaw2, ckchong2}@live.utm.my, akmalmuhd@gmail.com

[2] Institute for Artificial Intelligence and Big Data, Universiti Malaysia Kelantan, City Campus, Pengkalan Chepa, 16100 Kota Bharu, Kelantan, Malaysia

[3] Faculty of Bioengineering and Technology, Universiti Malaysia Kelantan, Jeli Campus, Lock Bag 100, 17600 Jeli, Kelantan, Malaysia
{saberi, safaai}@umk.edu.my

[4] Soft Computing and Intelligent System Research Group, Faculty of Computer Systems and Software Engineering, Universiti Malaysia Pahang, 26300 Kuantan, Pahang, Malaysia
arfian@ump.edu.my

[5] Biomedical Research Institute of Salamanca/BISITE Research Group, University of Salamanca, Salamanca, Spain
corchado@usal.es

[6] Department of Electronics, Information and Communication Engineering, Osaka Institute of Technology, 535-8585 Osaka, Japan
omatu@rsh.oit.ac.jp

Abstract. Mathematical models of metabolic processes are the cornerstone of computational systems biology. In model building, the task of parameter estimation is difficult due to the huge numbers of kinetics parameters involved. The common way of estimating the parameters is to formulate it as an optimization problem. Global optimization methods can be applied by minimizing the distance between experimental data and predicted models. This paper proposes the Hybrid of Bees Algorithm and Harmony Search (BAHS) to estimate the kinetics parameters of essential amino acid production in the aspartate metabolism for *Arabidopsis thaliana*. The performance of the BAHS is evaluated and compared with other algorithms. The results show that BAHS performed better as it improved the performance of the original BA by 60%. Meanwhile, it takes less computational time to estimate the kinetics parameters of essential amino acid production for *Arabidopsis thaliana*.

Keywords: Systems biology · Parameter estimation · Bees Algorithm
Harmony Search · Computational intelligence · *Arabidopsis thaliana*

© Springer Nature Switzerland AG 2019
F. Fdez-Riverola et al. (Eds.): PACBB 2018, AISC 803, pp. 9–16, 2019.
https://doi.org/10.1007/978-3-319-98702-6_2

1 Introduction

Mathematical modelling has become a fundamental tool for systems biology to better understand the regulation and dynamic behaviour of metabolism. Moreover, it gives an understanding and prediction of the simulation of biological and experimental processes. There are nine operations in mathematical modelling, which include a selection of data, information collection on the network structure and regulation, assumption and simplification, selection of the framework for mathematical modelling, estimation of parameters, model diagnostics, validation of the model, refinement and application of the model [1]. Biological processes models are usually represented by mathematical expressions which depend heavily on the ordinary differential equations (ODEs). Meanwhile, these mathematical expressions also refer to the mathematical modelling, and their measurements of system kinetics parameters include the rate of reaction, production and decay coefficient, and approximation or reduction which are satisfied by the systems dynamic. Basically, the measurement of these parameters is usually hard, and at the same time, the nonlinearity and intricacy of the biological processes have given rise to significant challenges [2].

Moreover, the huge numbers of kinetics parameters involved in the metabolism also make the wet laboratory process more complex, costly and time-consuming, in addition to which they are usually unknown and difficult to obtain experimentally. The incompleteness of the available experimental results causes difficulty in finding the optimum values. These limitations may affect the plausible parameters that represent the actual biological processes. Parameter estimation is one of the key steps in the modelling as it helps to characterize the system once the parameter value is estimated completely and accurately. It acts as an optimization to achieve the optimum values and best fitting with the corresponding experimental analyses, and this procedure is used in determining suitable numerical parameters that convert the symbolic model into a numerical model and make it fit the experimental results [1]. Among the nine operations in mathematical modelling, parameter estimation is the most crucial and challenging step, as the operations in phases preceding parameter estimation do affect the difficulties in estimation.

A common way of estimating the parameters is to formulate parameter estimation problem as an optimization problem, which consist of finding the parameters that give the best fit to a set of experimental data. To solve this problem, stochastic global optimization algorithms can be used, such as a Genetic Algorithm (GA), Simplex Downhill (SD), Simulated Annealing (SA), Bees Algorithm (BA), Harmony Search and many more. GA is a popular algorithm that is useful for solving problems of nonlinearity and does not require a constant initiation value for parameter estimation. However, it is easily trapped in local minima and thus a longer time is needed to search for the optimum [3]. The other algorithms also have limitations, as SA takes a longer time to achieve convergence in order to search for the optimum [3] and SD has a limited speed of convergence when raising the merit space's dimensions [4].

In this research, a hybrid of the Bees Algorithm and Harmony Search (BAHS) which is implemented in SBToolbox for MATLAB [5] is proposed to estimate the kinetic parameters of the aspartate metabolism in *Arabidopsis thaliana*. Essential amino acids are amino acids that are unable to be synthesized in the human body, such that humans need to obtain them by consuming them from other sources or from animal protein. This has the effect that it is not possible to study and understand the production of essential amino acids in the human body. As an alternative to giving better understanding of this model, *Arabidopsis thaliana* is selected because this plant is able to produce both essential and non-essential amino acids. This research has significance in modelling the production of essential amino acids in the aspartate metabolism of the model plant *Arabidopsis thaliana* and estimating parameter values by using the proposed method. Moreover, a performance evaluation comparison of BAHS is also conducted in this study together with the existing algorithms.

This paper is organized as follows: In Sect. 2, materials and methods are briefly discussed. Section 3 is mainly about the experimental results and discussion. Lastly, the paper is summarized in the concluding section and future work is suggested.

2 Materials and Methods

2.1 Datasets and Experimental Setup

The aspartate metabolism of model plant Arabidopsis thaliana [6] is used in this research. It is also known as the aspartate-derived amino acid pathway. This pathway is responsible for the production of lysine (Lys), threonine (Thr), methionine (Met) and isoleucine (Ile) which are the essential amino acids synthesized from the carbon flux of aspartate (Asp). In this research, the kinetic parameters for Lys, Thr, and Ile were estimated using proposed BAHS in SBToolbox. The kinetic parameters for each of the essential amino acids are as follows: Ile with 6 parameters, Lys with 9 parameters and Thr with 16 parameters. The experiment was conducted using MATLAB R2010a, SBToolbox (version 157 2.0.5), COPASI (version 4.8), with Intel i3 processor and 4 GB RAM.

2.2 A Hybrid Bees Algorithm and Harmony Search (BAHS)

Previously BA and HS were implemented separately in solving the parameter estimation [7, 8]. Hence, the hybrid Bees Algorithm and Harmony Search (BAHS) is proposed in this study. BAHS integrates two optimization algorithms to perform the parameter estimation. Each of the optimization algorithms can cover the weakness of the other. The Bees Algorithm has the weakness that premature convergence may occur as it always searches for the global optimum while ignoring the local optima between solutions in a population, while the Harmony Search algorithm can balance the intensification and diversification. Pitch adjustment and random selection regulate the diversification (exploration) to retain good local optima. The intensification (exploitation) in the Harmony Search algorithm is regulated by harmony memory consideration, which allows the process of searching towards good solutions for the searching space. Figure 1 illustrate the pseudocode and flowchart for BAHS. Previously, the original BA only involved 3 phases, but after hybridizing with HS, there are 6 main phases involved. The phases are as follows:

2.2.1   Initiate Population with Random Solutions

This starts with the Bees Algorithm (BA) where n scout bees are located in the search space simultaneously. Next, the scout bees evaluate the fitness of each site that has been visited. Then if the stopping criterion is not met, the bees with the highest fitness are called "selected bees" and the sites they visited are selected for neighbourhood search.

2.2.2   Neighbourhood Search

The next steps are the neighbourhood search of the selected site, where more bees are assigned to find the best e site. There is a restriction where it needs to decrease the number of points to be explored, in which case only those bees with the highest fitness in each patch are chosen to form a new population. The rest of the bees $(n - m)$ in the population are assigned simultaneously to surround the search space, scouting for new potential solutions and then evaluating their fitness.

2.2.3   Initiate Harmony Memory

Here, the Harmony Search (HS) algorithm is added and performs a stochastic random search instead of a gradient search to avoid getting stuck in the local minima. The HS is iteratively improved and initiates a population of random harmonies that are located in the Harmony Memory (HM) and also initiates the parameters mentioned.

2.2.4   Generate New Harmony

During each repetition, which is also the generation of a new harmony step, the latest harmony is created depending on the three rules mentioned, which are the Harmony Memory Considering Rate (HMCR) used to choose the variables of new harmonies from overall Harmony Memory (HM). Suitable values for HMCR are in the range between 0 and 1 in order to find the global optimum. An HMCR of 0.90 means that the algorithm selects a variable value from the HM with a probability of 90% at the next step. Then, the Pitch Adjusting Rate (PAR) is responsible for local improvement by escaping from local optima, where a value of the PAR set to be 0.10 means that the algorithm selects a value for neighbouring search with 10% probability, with an upper value of 5% or a lower value of 5%. Lastly, the Random Selection (RS) is used to provide random elements for the new harmony in this phase.

2.2.5   Update HM

The latest harmony is better; it then evaluates the fitness and updates the worst harmony in the HM. The objective function is minimized in the proposed algorithm for parameter estimation, and is expressed as below:

$$F(X) = \min \arg \sum_{m=1}^{M} \sum_{n=1}^{N} (\hat{y}_n - y_n(x_m))^2 \tag{1}$$

where $\hat{y}_n$ represents the value of experimental output at n time point, y_n is the value of simulated output at n time point, x is the solution corresponding to the parameters set, M represents the total number of parameters to be estimated and N represents the simulated times [9].

2.2.6   Termination

The process is iterated until a stopping condition is met, which based on the Number of Improvisations (NI) or known as the number of iterations.

```
Hybrid of Bees Algorithm and Harmony Search (BAHS)
Start: Initialize population with randomsolutions;
Evaluate fitness of population
   While (termination not met)
   New population formed;
   Choose sites for neighbourhood search;
   Recruit bees for selected site;
   Evaluate fitnesses;
   Choose the fittest bee from each patch;
   Assign the rest of the bees to search at random;
   Evaluate their fitnesses.
Initialized Harmony Memory (HM)
   For I=1 to number of decision (N)
   do
   R1= uniform random number between 0 and 1
   If (R1<P_HMRC) //memory consideration
        X[I] will be selected from harmony search randomly
        R2 = uniform random number
   If (R2<P_PAR) //pitch adjustment
        X[I]=X[1]± Δ
   End if
        Else if //random selection
        X[I]=X∈ φ (φ=value set)
        End If
End do
fitness_X=evaluate_fitness(X) //evaluate the fitness of each vector
Update_memory(X, fitness_X) //update harmony memory if applicable
End while
Check:
If termination == true;
End;
```

Note: The highlighted box is the Harmony Search (HS) that is hybridized into Bees Algorithm (BA).

Fig. 1. The pseudo code of BAHS

3   Results and Discussion

In this study, time series data for the concentration of essential amino acids: isoleucine, lysine and threonine inside the aspartate metabolism were generated to evaluate the accuracy of each estimation algorithm. According to the time series data obtained, the values of computational time, average error rate (A) and standard deviation (STD) are recorded, calculated and used as the performance measurements for BAHS and compared to the four other estimation algorithms, BA, GA, SD and SA.

Table 1 presents the comparison in terms of the computational time, average error rate (A), and standard deviation (STD) for isoleucine, lysine and threonine. The table shows that the computational time, average error rate and standard deviation yielded by BAHS proved to be the smallest compared to BA, GA, SD and SA. These indicate that BAHS performed better and faster, as the standard deviation value is the closest to 0 with the lowest average error rate and computational time. BAHS was shown also to be closer to the experimental line, with the smallest difference of any of the simulated line, as in Fig. 2.

Table 1. Comparison between BAHS, GA, SD and SA for isoleucine, lysine and threonine in terms of computational time, average error rate, and standard deviation.

Data	Measurements	Algorithms				
		BAHS	BA	GA	SD	SA
Isoleucine	Computational Time (seconds)	126.0289	157.8965	227.9899	130.5578	777.9994
	Average error rate, A	0.000383	0.001795	0.00057	0.000761	0.001153
	Standard deviation, STD	0.000443	0.003181	0.000812	0.000448	0.001989
Lysine	Computational Time (seconds)	174.0265	187.1365	318.4473	376.5896	1518.0464
	Average error rate, A	0.011821	0.014949	0.276142	0.084045	0.040609
	Standard deviation, STD	0.018405	0.028613	0.311897	0.099833	0.034668
Threonine	Computational Time (seconds)	245.9602	264.4962	388.0422	362.3243	1794.907
	Average error rate, A	0.00369	0.016398	0.086734	0.011982	0.006555
	Standard deviation, STD	0.007893	0.02318	0.169958	0.016982	0.007895

*Shaded column indicates the best results

Overall, the results show that BAHS outperformed BA, GA, SD, and SA for each of the essential amino acids. The computational time taken to estimate the kinetics parameters is higher among other algorithms. In this regard, BAHS obtained slight reduction on computation time compared to others. BAHS is introduced as a hybrid of the Bees Algorithm with Harmony Search and therefore enjoys the advantages of both in searching for the optimum minima by minimizing the objective function. This is because BAHS includes the global optimum search by BA while HS addresses the limitation of BA that it may lead to premature convergence as it always searches for the global optimum and ignores local optima [10]. BAHS has balanced the intensification and diversification in a way that retains good local optima by the pitch adjustment operation from HS along with random selection and memory consideration, all acting to search for a good solution [11].

The performance of SA was the slowest of all the algorithms for each of the essential amino acids, as its parallelism is not easy to exploit and it needs a long time for convergence in order to search for the best optimum [3, 12]. The performance of GA was also slow to estimate the kinetic parameters, as it is easily trapped in local minima and its operation in the algorithm is more complex than BAHS, having selection, mutation and crossover operations in its algorithm [3, 13]. SD performs with a greatly limited speed of convergence when its merit space dimension is increasing so that more time is required to find the optimum value [4].

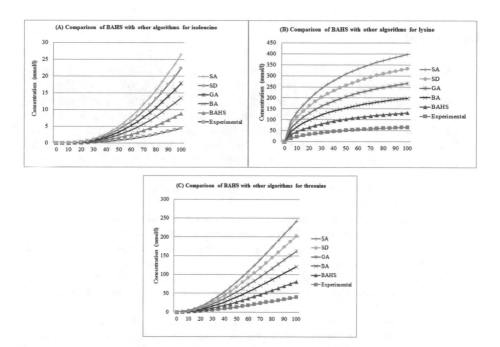

Fig. 2. Comparison of experimental and simulated lines of BAHS, BA, GA, SD and SA for isoleucine (A), lysine (B) and threonine (C).

4   Conclusion

In conclusion, the performance of BAHS for parameter estimation in this work is better than BA, GA, SD, and SA. According to the findings, the results obtained are more consistent in BAHS, as the value of standard deviation is closer to zero and with a very low error rate compared to the other algorithms. Furthermore, the model simulation shows that the simulated curve line of BAHS is closest to the experimental line. Additionally, BAHS also took less computational time than the other algorithms. This is because the hybrid of Bees Algorithm and Harmony Search has covered the limitation in BA. The hybrid algorithm enhanced the search capability by giving good balances between intensification and diversification by means of pitch adjustment in Harmony Search. In future work, a comparison may be conducted and tested with various artificial noise in order to determine how noise influences the parameter estimation, which act as the best way to assess the performance of various optimization algorithms.

Acknowledgements. We would like to thank Malaysian Ministry of Higher Education and Universiti Teknologi Malaysia for supporting this research by the Fundamental Research Grant Schemes (grant number: R.J130000.7828.4F886 and R.J130000.7828.4F720). We would also like to thank Universiti Malaysia Pahang for sponsoring this research via the RDU Grant (Grant Number: RDU180307).

References

1. Chou, I.C., Voit, E.O.: Recent developments in parameter estimation and structure identification of biochemical and genomic systems. Math. Biosci. **219**, 57–83 (2009)
2. Remli, M.A., Deris, S., Mohamad, M.S., Omatu, S., Corchado, J.M.: An enhanced scatter search with combined opposition-based learning for parameter estimation in large-scale kinetic models of biochemical systems. Eng. Appl. Artif. Intell. **62**, 164–180 (2017)
3. Baker, S.M., Schallau, K., Junker, B.H.: Comparison of different algorithms for simultaneous estimation of multiple parameters in kinetic metabolic models. J. Integr. Bioinform. **7**, 1–9 (2010)
4. Koshel, R.J.: Enhancement of the downhill simplex method of optimization. Proc. SPIE **4832**, 270–282 (2002)
5. Schmidt, H., Jirstrand, M.: Systems Biology Toolbox for MATLAB: a computational platform for research in systems biology. Bioinformatics **22**, 514–515 (2006)
6. Curien, G., Bastien, O.: Understanding the regulation of aspartate metabolism using a model based on measured kinetic parameters. Mol. Syst. Biol. **5**, 271 (2009)
7. Leong, Y.Y., Chong, C.K., Choon, Y.W., En, L., Deris, S., Illias, R.M., Omatu, S., Saberi, M.: Simulation of fermentation pathway using Bees Algorithm. Adv. Distrib. Comput. Artif. Intell. J. **1**, 13–19 (2013)
8. Bahamish, H.A.A., Abdullah, R., Salam, R.A.: Protein conformational search using Bees Algorithm (2008). http://ieeexplore.ieee.org/lpdocs/epic03/wrapper.htm?arnumber=4530597
9. Tashkova, K., Korošec, P., Šilc, J., Todorovski, L., Džeroski, S.: Parameter estimation with bio-inspired meta-heuristic optimization: modeling the dynamics of endocytosis. BMC Syst. Biol. **5**, 159 (2011)
10. Pham, D.T., Castellani, M.: The Bees Algorithm: modelling foraging behaviour to solve continuous optimization problems. Proc. Inst. Mech. Eng. Part C J. Mech. Eng. Sci. **223**, 2919–2938 (2009)
11. Nguyen, K., Nguyen, P., Tran, N.: A hybrid algorithm of Harmony Search and Bees Algorithm for a University Course Timetabling Problem. Int. J. Comput. Sci. Issues **9**, 12–17 (2012)
12. Wang, Z.G., Wong, Y.S., Rahman, M.: Optimisation of multi-pass milling using genetic algorithm and genetic simulated annealing. Int. J. Adv. Manuf. Technol. **24**, 727–732 (2004)
13. Elbeltagi, E., Hegazy, T., Grierson, D.: Comparison among five evolutionary-based optimization algorithms. Adv. Eng. Inform. **19**, 43–53 (2005)

In Silico Modeling and Simulation Approach for Apoptosis Caspase Pathways

Pedro Pablo González-Pérez[1(✉)] and Maura Cárdenas-García[2]

[1] Universidad Autónoma Metropolitana,
Av. Vasco de Quiroga 4871, 05300 Ciudad de México, Mexico
pgonzalez@correo.cua.uam.mx
[2] Benemérita Universidad Autónoma de Puebla,
13 Sur 2702, Col. Volcanes, 72420 Puebla, Mexico

Abstract. We revisit and improve *in silico* modeling and simulation approach of the apoptosis caspases pathways, initially developed for exploring and discovering the complex interaction patterns of apoptotic caspases and the mitochondrial role. Symbolic abstractions and algorithms of the *in silico* model were improved to allow dealing with crucial aspects of the cellular signal transduction such as cellular processes. Also, the particular model of extrinsic and intrinsic apoptotic signaling pathways was improved, increasing the number of reactions and using all kinetic parameters accurately calculated. Using the computational simulation tool BTSSOC-Cellulat, we were able to determine experimentally how the modulation of concentrations of proteins XIAP, cFLIPs and TRAIL/ FASL, can cause the death of cancerous cells. Our results show how crucial were the improvements made in the *in silico* modeling approach, which in turn were reflected in the accuracy of the simulation and, therefore, in the significant value of the *in silico* experiments carried out.

Keywords: Caspase signaling pathway · Apoptosis
In silico modeling and simulation approach · *In silico* experiments

1 Introduction

All the organisms can eliminate cells that are no longer necessary, damaged or infected, developing a programmed mechanism of death called apoptosis [1]. Apoptosis is characterized by a series of biochemical and morphological changes [2]. There are other types of cellular death, sometimes with characteristics similar to apoptosis but that do not adjust completely to it. Caspases are Cysteine Dependent Aspartate Specific Proteases in a family closely related to C14 or clan CD, were described for the first time 20 years ago. Caspases participate in the control of programmed cell death or apoptosis, the inflammatory response, proliferation and cellular differentiation. These enzymes have been intensively studied due to their participation in the development of illnesses such as cancer, neurodegenerative diseases or autoimmune diseases [3].

The initiator caspases are caspase-2, -8, -9 and -10 and these are the first ones to be activated after an apoptotic stimulus. The effector caspases are caspase-3, -6 and -7, activated by the initiators and are in charge to degrade different cellular substrates

© Springer Nature Switzerland AG 2019
F. Fdez-Riverola et al. (Eds.): PACBB 2018, AISC 803, pp. 17–26, 2019.
https://doi.org/10.1007/978-3-319-98702-6_3

creating the characteristic changes of the apoptosis [4]. There are two different signaling pathways that lead to the activation of apoptosis: (1) the pathway of the death receptors or extrinsic, that implicates the activation of determined members of the superfamily of receptors of the tumor necrosis factor, and (2) the mitochondrial pathway or intrinsic, induced by distinct forms of cellular stress, caused by deprivation of growth factors, DNA damage, adhesion lost, ER stress, ionizing radiation, UV radiation, etc. [5]. Each signaling pathway produces the activation of the initiator caspases and leads to the activation of the executor caspases which, if regulated adequately will lead a cell to death by apoptosis or by necroptosis [6].

Experimental research in cellular and molecular biology has found an invaluable support from the broad spectrum of *in silico* modeling and simulation approaches developed in the last few years. Specifically, *in vitro* studies involving cell signal transduction systems have been strongly favored from a wide range of *in silico* computational tools proposed and available [7]. Opportunely, many of these tools are in themselves robust virtual laboratories to conduct *in silico* experiments, that are able to complement, optimize and feedback the *in vitro* counterpart [8–12].

In this work, we revisit, improve and extend an *in silico* modeling and simulation approach for cellular signal transduction systems and its former application in the inspection and prediction of the apoptosis caspases pathways. We model both intrinsic and extrinsic pathways of apoptosis, including the proteins of the Bcl-2 family with pro and anti-apoptotic activity [13], with the aim of identifying, by *in silico* experiments, regulation points avoided by cancer cells that allow them to survive. The *in silico* experiments showed how, at a molecular level, cancer cells evade the signaling pathway of caspases, resisting mitochondrial induction. The revisited *in silico* approach was integrated in BTSSOC-Cellulat [14, 15], a computational simulation tool for cellular signaling networks developed by us a few years ago.

2   Materials and Methods

2.1   The *in Silico* Modeling and Simulation of Cellular Signal Transduction

The signal transduction model presented here is itself an improvement and extension of the model implemented by the computational simulation tool BTSSOC-Cellulat [14, 15], developed and released by our working group in recent years. The aim of this process of improvement and extension was (1) in addition to chemical reactions, also to incorporate into the model the cellular processes that arise in the cell as a result of particular configurations of activation/inhibition of the signaling molecules, which lead to certain cellular states, and (2) to refine crucial aspects of the model such as multi-compartmentalization, location, topology and timing.

Both the original model and the new improved model are based on the concepts of tuple and tuple space [16] for the representation of cellular structures and signaling components. That is, a cell C_i is conceived as a set of tuple spaces given by expression (1), where each tuple space TS_{ij} models a specific cellular compartment, as shown in Fig. 1. On the other hand, the tuples are used as logical abstractions that model the

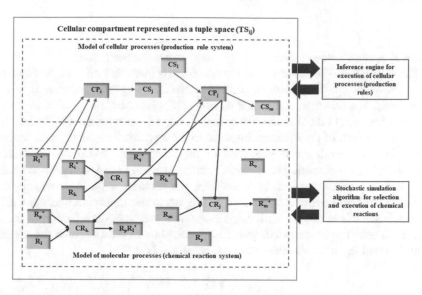

Fig. 1. A cellular compartment modelled as a tuple space. Reactions, cellular processes and reactants are modelled as tuples.

three types of signaling components considered in the model, as defined by expression (2), and are referred to chemical reactions, cellular processes and reactants, as described by expressions (3), (4) and (5), respectively.

$$C_i = \{TS_{i1}, TS_{i2}, \ldots, TS_{in}\} \tag{1}$$

where C_i, $1 \leq i \leq m$, the *i-th* cell belonging to the tissue or cellular group G and TS_{ij}, $1 \leq j \leq n$, is a set of tuples.

$$\forall t \in TS_{ij}, \ 1 \leq j \leq n, \ t = cr, \ t = cp, \ t = r, \ or \ t = p \tag{2}$$

where cr denotes a reaction, cp a cellular process, r a reactant, and p a product.

$$cr([(r_1, reqMol_1), (r_2, reqMol_2)], K, [(p_1, pm_1), (p_2, pm_2)]) \tag{3}$$

where: r_1, r_2 are reactants and $reqMol_1$, $reqMol_2$ are the number of reactant molecules r_1, r_2, respectively; K is the reaction rate constant; p_1, p_2 are products and pm_1, pm_2 are the number of molecules formed of products p_1, p_2, respectively.

$$cp([cond_1, \ldots, cond_p], [act_1, \ldots, act_q]) \tag{4}$$

where: $cond_1$, ..., $cond_p$ represent the particular configuration of activation/inactivation of certain signaling components required to obtain the cellular state and molecular actions represented by act_1, ..., act_q. Note that expression (4) is the logical abstraction for production rules of type IF <condition> THEN <action> .

$$r(r_i, Mol_i) \tag{5}$$

where: r_i denotes reactant I and Mol_i its micro molar concentration.

The simulation of cellular signal transduction requires that all reactions and reactants that make up the system have been modeled using expressions (3) and (5), respectively. The selection and execution of a particular chemical reaction depends on two main factors: (1) the reaction rate constant (value K in expression (3)), and (2) the availability of each of the reactants involved in the reaction. The control mechanism in charge for the selection and execution of the reactions is based on the Gillespie algorithm [17], a stochastic simulation algorithm typically used to mimic systems of chemical/biochemical reactions in an efficient and accurate way. Expressions (6), (7) and (8) listed below represent the core of this stochastic algorithm and are used for (1) calculation of the rate for each eligible chemical reaction, (2) selection of the next chemical reaction to be executed, and (3) determination of time between the execution of the last and the next chemical reaction, respectively.

$$Rate = K * \prod_{i=1}^{2} \left(\frac{Mol_i}{reqMol_i} \right) \tag{6}$$

$$\psi \le \frac{\sum_{i=1}^{k} Rate_i}{RTot} \tag{7}$$

where $RTot$ is the summation of the rates of all eligible reactions.

$$Stop_{time} = \frac{-\ln(\tau)}{RTot} \tag{8}$$

On the other hand, cellular processes are modeled as production rules and represented as tuples formulated by expression (4). The execution of these production rules is carried out by an inference engine that applies logical rules to the set of production rules to produce new facts, i.e., new tuples representing cellular events or processes, as illustrated in the top view of Fig. 1.

The improvements of the model of cellular signal transduction previously introduced were integrated as an update of the BTSSOC-Cellulat computational simulation tool developed by our working group a few years ago. BTSSOC-Cellulat (http://bioinformatics.cua.uam.mx/node/10) is an integrated virtual environment for *in silico* experimentation on inter and intracellular signaling networks, characterized by multi-compartmentalization, location, topology and timing.

2.2 The Caspase Signaling Model

Caspases constitute a family of enzymes very common in evolution; they are central components of the apoptosis induction machinery. The group of the initiator caspases is formed by caspase-2, -8, -9 and -10, and these are the first to be activated after an apoptotic stimulus. This stimulus causes the assembly of activation complexes, which are made of the inactive forms of the caspases themselves and adaptive proteins, and constitute a platform for the activation of the initiator caspases. As previously mentioned, there are two general types of signaling pathways that lead to activation of programmed cell death: the intrinsic and extrinsic pathways. Both pathways produce the activation of the initiator caspases and converge on the activation of the executor caspases responsible of the characteristics of apoptosis. Above 50% of neoplasms present defects in apoptotic machinery. Among the better characterized effects are the increase in the expression of some anti-apoptotic proteins of the Bcl-2 family and mutations in the tumor suppressor gene TP53, which codifies for the pro-apoptotic protein p53. Figure 2 illustrates in a simplified way the interactions that characterize the extrinsic and intrinsic pathways. Such interactions, if properly regulated, cause a cell to undergo death by apoptosis or necroptosis.

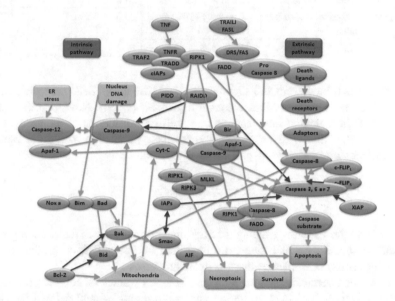

Fig. 2. A simplified view of the interactions that are observed in the extrinsic and intrinsic apoptosis signaling pathways [5, 6, 18, 19]. In this signaling model proteins are represented as solid blue ellipses, cellular events as solid orange rectangles, and organelles involved as solid green. Red arrows indicate inhibition relationships and green arrows indicate activation relationships.

3 Results and Discussion

3.1 From Modeling to Simulation of Caspase Signaling Pathways

The reactions and reactants included in the simulation are initially identified and specified from the caspase signaling network illustrated in Fig. 2. That is, each arc (directed edge) joining two nodes constitutes a reaction, while the nodes represent the reactants. The Table 1 provides a summary of these specified reactions, as part of the modeling phase. The kinetic parameters of the reactions and concentration values of the

Table 1. A summary of signaling elements –reactions and reactants– defined as part of the modeling of caspases signaling pathways.

Reaction	Concentration (µM)	Km (µM)	Vmax (µM/mg/min)	V₀
Caspase9* + APAF-1* -> Caspase/APAF	APAF-1 (6.8)	42.0	2.8	0.390
Caspase/APAF + Caspase6 -> Caspase6*	Caspase6 (1.8)	31.0	19.0	1.042
Caspase9* + Bir -> Caspase9	Bir (2.7)	6.7	55.0	15.797
Caspase8* + cFLIP + Caspase7 -> Caspase7*	cFLIPL (22.0)	9.0	16.0	3.692
Cyt-C + Caspase3 -> Caspase3*	Caspase3 (5.6)	26.0	11.7	2.073
Caspase3* + Caspase_Substrate -> Caspase_Substrate*	Capase_Substrate (5.0)	24.6	15.6	2.635
IAPs* + Caspase6* -> Caspase6	Caspase6 (1.8)	13.1	2.5	0.302

reactants are obtained from the literature [20, 21] or calculated, the value of V_0 is obtained from the Michalis-Menten equation (expression (9)). Note, that the reactions are expressed in a natural language, as they will be introduced in the computational simulation tool, which will be responsible for their translation to the corresponding logical representation structures previously discussed.

$$V_0 = \frac{V_{max} * [S]}{K_m + [S]} \tag{9}$$

Also, from the caspase signaling network illustrated in Fig. 2, the cellular processes that lead to specific cellular states are identified and specified as production rules. Table 2 lists some examples of these cellular processes.

Once all the chemical reactions with their kinetic parameters, cellular processes and reactants with their initial concentration values are identified and defined (see a summary in Tables 1 and 2), the next step is the recording of all these signaling elements in the computer simulation tool BTSSOC-Cellulat, with which the simulation of caspase signaling pathways will be ready to start with *in silico* experiments. As a result, the simulated caspases signaling network is made up of 43 nodes representing reactants, 3 nodes representing cell processes – such as survival, apoptosis and necroptosis – and 35 arcs representing reactions between the involved nodes. The cellular compartments involved in the simulation are extracellular space, cell membrane, cytosol and mitochondria.

3.2   *In Silico* Studies of the Caspases Signaling Pathway

The highest goal of the *in silico* studies was to identify the critical combination of target signaling elements and the required concentrations, that lead the cell to death. To achieve the above, the following phases of *in silico* experiments were carried out: (1) the simulation was run from the initial concentration values of the reactants

Table 2. Examples of cellular processes defined as part of the modeling of extrinsic intrinsic caspases signaling pathways.

Cellular process	Production rule
Apoptosis (RULE #1)	IF active_conc("Caspase_substrate*", AC_Caspase_substrate*) AND AC_Caspase_substrate* > Th_Caspase_substrate AND active_conc("AIF*", AC_AIF*) AND AC_AIF* > Th_AIF THEN triggered_cellular_event("APOPTOSIS") AND stop_All_Reactions(_).
Necroptosis (RULE #2)	IF available_conc("RIPK1/MLKL/RIPK3", AvC_ RIPK1/MLKL/RIPK3) AND AvC_ RIPK1/MLKL/RIPK3 > Th_RIPK1/MLKL/RIPK3 THEN triggered_cellular_event("NECROPTOSIS") AND stop_All_Reactions(_).

Table 3. Combinations of micro molar concentrations for the target signaling elements.

Target signaling elements	Experiments/concentration (μM)			
	No. 1	No. 2	No.3	No. 4
Caspase 8	4.2	42	420	42
Caspase 9	7.4	74	740	74
XIAP	4.3	0.43	0.43	0.043
cFLIPs	19	1.9	1.9	0.19

involved (see some examples in Table 1). When necessary, the kinetic parameters were adjusted until the simulated signaling network was able to reproduce the expected results for this configuration of initial concentrations, (2) a switchable "on-off" technique was applied over the caspase signaling network to detect the critical combination of target signaling elements responsible for carrying the cell to death, and (3) new *in silico* experiments were designed increasing/decreasing the concentration of the target signaling elements previously identified in magnitude of 10.

During the development of the in-silico experiments, we modified the concentrations of the different proteins, belonging to both the intrinsic and extrinsic pathways, as can be seen in Table 3. What we should observe is if decreasing the concentration of cFLIPs and XIAP, keeping the concentration of Caspase-8 constant, and increasing the concentration of Caspase-9, the cell undergoes an apoptosis.

As we modify the concentration of two initiator caspases (-8 and -9) and increase the concentration of the inhibitor proteins XIAP and cFLIPs, we observe that the cell survives. Nevertheless, if we decrease the concentration of XIAP and cFLIPs, the cell dies by apoptosis, as shown in Fig. 3. In a relevant way, we observed that by increasing

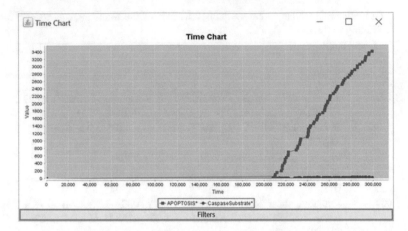

Fig. 3. Apoptosis (red solid squares). Thanks to this *in silico* experiment we were able to determine that by modifying the concentrations of caspases-8 and -9 and decreasing the concentration of XIAP and cFLIPs, the cell undergoes apoptosis. On the chart, the x-axis represents the time in milliseconds, and the y-axis represents the concentration of reactants (scaled by 10) in micromolar.

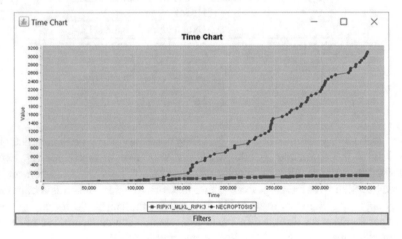

Fig. 4. Necroptosis (blue solid circles). Thanks to this *in silico* experiment we were able to determine that by increasing the concentrations of APAF-1 the cell undergoes necroptosis. On the chart, the x-axis represents the time in milliseconds, and the y-axis represents the concentration of reactants (scaled by 10) in micromolar.

the concentration of APAF-1 the cell dies due to necroptosis, as can be seen in Fig. 4. On the other hand, by increasing the concentration of TRAIL/FASL the cell survives, independently of the damage to the DNA, ER stress and other apoptotic signals (not shown).

4 Conclusions

Caspases participate in the control of cell death by apoptosis, inflammatory response, proliferation and cellular differentiation. In the last years, there has been an increase of information at the molecular level of its regulation, as caspases are involved in the development of cancer, autoimmune diseases and neurodegenerative diseases. Therefore, our caspase signaling model was increased with a considerable number of new interactions. On the other hand, our computational simulation tool has also been improved, allowing us not only to model the reactions involved in the extrinsic and intrinsic apoptotic signaling pathways under different conditions, as we did previously [18, 19]. Now we can also (1) model the cellular processes that are triggered from particular configurations of activation/inactivation of signaling molecules, and their impact on the resulting cellular state (as indicated in expression (4) and illustrated in Table 2), (2) calculate accurately the kinetic parameters of the reactions required by the simulation tool, based on those reported in the specialized literature, and (3) simulate the behavior of complex cellular signaling networks, considering the role of key aspects such as topology, location and timing, which were improved.

Throughout in silico experiments, we could modify the initial concentrations of the proteins which participate in extrinsic and intrinsic apoptotic signaling pathways, and modulate them to cause the cell death. Specifically, with the support of in silico experimentation, we observed that the possible control proteins are XIAP, cFLIPs and TRAIL/FASL. This will allow us to carry out *in vitro* experiments, testing the concentrations suggested by *in silico* experiments, focusing on triple-negative breast cancer which is hormone-independent and thus one of the hardest to treat.

References

1. Kerr, J.F.T., Wyllie, A.H., Currie, A.R.: Apoptosis: a basic biological phenomenon with wide-ranging implication in tissue kinetics. Br. J. Cancer **26**(4), 239–257 (1972)
2. Wyllie, A.H.: Apoptosis: an overview. Br. Med. Bull. **53**(3), 451–465 (1997)
3. Poreba, M., Szalek, A., Kasperkiewicz, P., Rut, W., Salvesen, G.S., Drag, M.: Small molecule active site directed tools for studying human caspases. Chem. Rev. **115**(22), 12546–12629 (2015). https://doi.org/10.1021/acs.chemrev.5b00434
4. Danial, N.N., Korsmeyer, S.J.: Cell death: critical control points. Cell **116**(2), 205–219 (2004). https://doi.org/10.1016/S0092-8674(04)00046-7
5. Poreba, M., Strózyk, A., Salvesen, G.S., Drag, M.: Caspase substrates and inhibitors. Cold Spring Harb. Perspect. Biol. **5**(8), a008680 (2013). https://doi.org/10.1101/cshperspect.a008680
6. Songane, M., Khair, M., Saleh, M.: An update view on the function of caspases in inflammation and immunity. Semin. Cell Dev. Biol. (2018). https://doi.org/10.1016/j.semcdb.2018.01.001
7. Alves, R., Antunes, F., Salvador, A.: Tools for kinetic modeling of biochemical networks. Nat. Biotechnol. **24**(6), 667–672 (2006). https://doi.org/10.1038/nbt0606-667

8. Ciocchetta, F., Duguid, A., Guerriero, M.L.: A compartmental model of the cAMP/ PKA/MAPK pathway in bio-PEPA. In: Third Workshop on Membrane Computing and Biologically Inspired Process Calculi (MeCBIC) (2009). http://dx.doi.org/10.4204/EPTCS. 11.5

9. Kerr, R.A., Bartol, T.M., Kaminsky, B., Dittrich, M., Chang, J.C., Baden, S.B., Sejnowski, T.J., Stiles, J.R.: Fast Monte Carlo simulation methods for biological reaction-diffusion systems in solution and on surfaces. SIAM J. Sci. Comput. **30**(36), 3126–3149 (2008). https://doi.org/10.1137/070692017

10. Hoops, S., et al.: COPASI: a complex pathway simulator. Bio-informatics **22**(24), 3067–3074 (2006). https://doi.org/10.1093/bioinformatics/btl485

11. Cowan, A.E., Moraru, I.I., Schaff, J.C., Slepchenko, B.M., Loew, L.M.: Spatial modeling of cell signaling networks. Methods Cell Biol. **110**, 195–221 (2012). https://doi.org/10.1016/ B978-0-12-388403-9.00008-4

12. Swat, M., Thomas, G.L., Belmonte, J.M., Shirinifard, A., Hmeljak, D., Glazier, J.A.: Multi-scale modeling of tissues using CompuCell 3D. Methods Cell Biol. **110**, 325–366 (2012). https://doi.org/10.1016/b978-0-12-388403-9.00013-8

13. Martinou, J.C., Youle, R.J.: Mitochondria in apoptosis; Bcl-2 family members and mitochondrial dynamics. Dev. Cell **21**(1), 92–101 (2011). https://doi.org/10.1016/j.devcel. 2011.06.017

14. González-Pérez, P.P., Omicini, A., Sbaraglia, M.: A biochemically inspired coordination-based model for simulating intracellular signalling pathway. J. Simul. **27**(3), 216–226 (2013). https://doi.org/10.1057/jos.2012.28

15. Cárdenas-García, M., González-Pérez, P.P., Montagna, S., Cortés Sánchez, O., Caballero, E. H.: Modeling intercellular communication as a survival strategy of cancer cells: an in silico approach on a flexible bioinformatics framework. Bioinf. Biol. Insights **10**, 5–18 (2016). https://doi.org/10.4137/BBI.S38075

16. Gelernter, D.: Generative communication in Linda. ACM Trans. Program. Lang. Syst. **7**(1), 80–112 (1985). https://doi.org/10.1145/2363.2433

17. Gillespie, D.T.: Exact stochastic simulation of coupled chemical reactions. J. Phys. Chem. **81**(25), 2340–2361 (1977). https://doi.org/10.1021/j100540a008

18. Cárdenas-García, M., González-Pérez, P.P., Montagna, S.: bioinformatics. EMBnet.journal **18**(S18.B), 94–96 (2012). Special Issue NETTAB 2012 Workshop on "Integrated Bio-Search". http://dx.doi.org/10.14806/ej.18.B.563

19. Cárdenas-García, M., González-Pérez, P.P.: Applying the tuple space-based approach to the simulation of the caspases, an essential signalling pathway. J. Integr. Bioinf. **10**(1), 225. https://doi.org/10.2390/biecoll-jib-2013-225. ISSN 1613-4516

20. Kang, W., et al.: Structural and biochemical basis for the inhibition of cell death by APIP, a methionine salvage enzyme. Proc. Nat. Acad. Sci. **111**(1), E54–E61 (2014). https://doi.org/ 10.1073/pnas.1308768111

21. Karki, P., Lee, J., Shin, S.Y., Cho, B., Park, I.S.: Kinetic comparison of procaspase-3 and caspase-3. Arch. Biochem. Biophys. **442**(1), 125–132 (2005). https://doi.org/10.1016/j.abb. 2005.07.023

Molecular Dynamic Simulations Suggest that P152R Mutation Within MeCP2 Can Lead to Higher DNA Binding Affinity and Loss of Selective Binding to Methylated DNA

Dino Franklin[✉]

Faculty of Computing, Federal University of Uberlandia,
Uberlandia, MG 38400-902, Brazil
dinofranklin@ufu.br

Abstract. MECP2 gene mutations can cause Rett Syndrome (RTT) – the second largest cause of mental retardation in girls. Studies based on the Methyl-Binding Domain (MBD) of MeCP2 and DNA complex crystals revealed how the mutated residues in the contact surface within DNA affect their binding. On the other hand, other mutations whose residue is not directly involved in the binding, are also related to RTT. In this paper, Molecular Dynamics (MD) and Potential of Mean Force (PMF) were applied to investigate how the MeCP2 P152R mutation influences MBD binding to DNA. The results suggest that the P152R mutation leads to MeCP2 increasing binding affinity towards both methylated and non-methylated DNA and decreasing binding selectivity towards methylated DNA. This may help explain previous inconclusive experimental results relating to the role of the P152R mutation in protein/DNA interactions and subsequent effects in RTT.

Keywords: Molecular Dynamics · Epigenetics · DNA-binding
Rett Syndrome · MeCP2 · P152R mutation

1 Introduction

Rett syndrome (RTT) is a postnatal neurodevelopmental disorder that mainly occurs sporadically and affects about 1:10000 female births. There is no known cure for RTT, but RTT does not appear to cause irreversible damage to neurons [1], which lends weight to the idea that therapeutic approaches might be possible.

Most cases of RTT are related to a set of genetic mutations within the MECP2 gene [2, 3]. In general, these mutations may cause the interruption of protein expression or lead to modifications in the protein structure and, consequently, to improper protein function [2]. More than half of MECP2 missense mutations associated with RTT are within its Methyl-CpG-binding domain (MBD) [4].

In general terms most of the previous studies have analyzed the RTT mutations that are directly involved in the MBD binding to DNA and have shown that the mutations reduce the affinity of MeCP2 towards methylated DNA [5, 6]. However, other than for the T158M and P101H mutations, there are no studies that explain the possible

© Springer Nature Switzerland AG 2019
F. Fdez-Riverola et al. (Eds.): PACBB 2018, AISC 803, pp. 27–34, 2019.
https://doi.org/10.1007/978-3-319-98702-6_4

mechanisms of how RTT missense mutations, that are not involved in the direct protein-DNA contact, influence its binding to DNA.

This study investigates how the P152R mutation can affect its binding to DNA. Although the P152R mutation is localized in a non-structured region far away from the DNA contact area, it is the fourth most common missense mutation in MBD related to RTT [3]. The approach used to investigate this phenomenon is Molecular Dynamic (MD) simulations. MD is widely regarded as offering one of the most powerful and promising methods for understanding macromolecular interactions and have been successfully used for studying interactions between proteins, DNA, lipids and small ligands [7]. Umbrella sampling is the MD technique used to analyze how the MeCP2 P152R mutation affects MBD binding to methylated and non-methylated DNA. In this case, a potential of mean force (PMF) curve is calculated for the neighborhood of the conformational space explored during simulation [8].

The significance of understanding such effects is that alterations in binding may have a direct influence on subsequent symptoms of RTT and may even represent a potential drug target for modulating disease progression. It is expected that this work contribute to establish the basis for a set of testable hypotheses for experimental studies relating to MeCP2 binding studies.

2   Materials and Methods

2.1   Initial Complex Structure and PDB Editing

The original structural model used for this study is the 2.5 Å resolution X-ray structure of the MeCP2-MBD in complex with methylated DNA that was originally reported by Ho et al. [9]. The original structural model is obtainable from Protein Data Bank (PDB) (Id.: 3C2I. http://www.pdb.org/pdb/explore/explore.do?structureId=3c2i) The MeCP2-MBD P152R mutation and the substitution of the selenomethionine to methionine were implemented using Swiss-PDB viewer [10]. The substitution of the 5mC (bases 5 and 33) to a cytosine in the DNA model was implemented in LeaP module of AMBER v9 [11]. The resulting topologies that were obtained through this process were subsequently optimized through the solutes relaxation in solvent approach using Molecular Dynamics. In order to obtain an unbiased complexation, the MeCP2-MBD was manually moved to a position about 7.5 nm away from the DNA using Pymol [12]. This topology was edited using LeaP and used as starting points for the (MeCP2-DNA) unbiased complexation.

The MBD amino acid sequence of the protein domain used is:

```
(ASASPKQRRSIIRD)
91--------100-------110-------120-------130------140
RGPMYDDPTLPEGWTRKLKQRKSGRSAGKYDVYLINPQGKAFRSKVELIM
141-------150-------160
YFEKVGDTSLDPNDFDFTVTGR (GSPSRHHHHHH)
```

The amino acid sequences in parentheses belong to the MBD domain, but were not detected in the X-ray experiments of Ho et al. [9] and were not considered in this study. The bold P letter shows where is the P152R mutation in the chain and the numbers refer to the correspondent residue/nucleotide in the original PDB file. The bold C letters indicate the 5-methylcytosines of the methylated DNA.

The DNA sequences were:

```
Chain B                          Chain C
1--------10--------20            21-------30--------40
TCTGGAACGGAATTCTTCTA             ATAGAAGAATTCCGTTCCAG
```

2.2   Force Field Parameters and Partial Atomic Charges

The addition of solvent and counter-ions to the in silico model, as well as the assignment of partial atomic charges and the preparation of topologies was implemented in LeaP [11]. K+, Na+ and Cl- ions were added at a proportion of about 100 nM to balance the total charge in the systems and to mimic the general ion concentration in mammalian cells.

The protein parameters were set by the FF03 force field [13]. The parmbsc0 force field was used to set the DNA parameters. The force field library for 5mC (structure optimization and partial atomic charges) were calculated using Restrained ElectroStatic Potential (RESP) by R.E.D. Server (RED Server 2011 http://q4md-forcefieldtools.org/REDServer/). The parmchk module of AMBER was used to prepare the other general amber force field (GAFF) parameters. The information concerning the new residue, 5-methylcytosine (5mC), was saved in GAFF mol2 format and used as LeaP input [11].

2.3   MD Preparation and Analysis

The topology and coordinate files were obtained using LeaP [11]. Furthermore, the edited pdb file was loaded and the solvate box was created using the TIP3PBOX water model and the added ions. The unit cell size was set to be within the 8 nm range from the edge of the box. The AMBER coordinates and topology files were converted to GROMACS using acpype.py program (available at https://github.com/t-/acpype). MD simulations were conducted with the GROMACS package, version 4.5.5 [14].

The DNA-MBD unbiased complexation simulations were prepared with a system energy minimization approach followed by isothermal–isobaric (NPT) ensemble equilibration [14]. The minimization algorithm used was the Steepest Descents Minimization with flexible water for 1000 steps [14]. The NPT equilibration was performed by DNA and protein position restrained MD for 40 ps. 10 unbiased complexation simulations were performed during a maximum period of 30 ns for the systems with MBD and methylated-DNA: 8 times the final structural conformations converged to the X-ray crystal conformation; and 2 times the macromolecules remained unbound. For the former ones, the final complexes have binding sites similar to the one presented in the pdb file. Unbiased complexation simulations were followed by DNA backbone position restrained MD simulation for 50 ns, and 50 ns of non-restrained simulations.

For the umbrella sampling windows simulations, short NPT equilibration (20 ps) was performed with DNA and protein position restrained. For the NPT equilibration, the DNA and protein were kept separate from the solvent in different coupling baths using Berendsen weak coupling method [14]. For the umbrella sampling simulations of 3 ns, the DNA and the protein were restrained at their window positions by a harmonic restraint with a constant of 1000 kJ mol^{-1} nm^{-2}. The temperature was maintained at 310K using the Nosé-Hoover thermostat algorithm and the pressure was regulated using a Parrinello-Rahman barostat algorithm [14]. All simulations used short-range electrostatic interactions and Van der Waals force cutoff of 1.4 nm, with long-range interactions calculated using the Particle Mesh Ewald (PME) algorithm and the leap-frog MD integrator with steps of 0.002 ps. Bonds were constrained by the Linear Constraint Solver (LINCS) algorithm [14].

The Hydrogen bonds, the Solvent Accessibility Surface (SAS) and the representative structures were calculated using the default parameters via the g_rmsf, g_hbond, g_sas and g_cluster programs included in the GROMACS package [14].

In order to calculate the binding energy between the protein and the DNA, the umbrella simulations were combined to generate a continuous PMF curve using the Weight Histogram Analysis Method (WHAM) [8].

3   Results

3.1   Structural Fit

Visual inspection in Fig. 1 suggests that the wild-type and P152R MBD structures are similar. The main difference is the distortion to the α-helix caused by the R152.

Fig. 1. MD simulations representative complex structures of the wild-type MeCP2-MBD (green) and the P152R MeCP2-MBD (magenta) complexed with methylated DNA. The red areas on surface illustrate the protein contact with the met-DNA. The 5mC are in yellow sticks.

3.2   Contact Surface

The contact surface calculation results (in Table 1) show that the total mutated MBD contact surface area towards methylated DNA is almost twice as large as the contact area of the wild-type MBD (see also Fig. 1) (Fig. 2).

Table 1. The contact surface average and standard deviation between the methylated DNA and wild-type and P152R MeCP2 MBD during backbone DNA restrained MD.

Sample	Hydrophobic [Å^2]	Hydrophilic [Å^2]	Total [Å^2]
Wild-type MBD	37.7 ± 8.0	166.2 ± 24.8	204.0 ± 29.4
P152R MBD	91.8 ± 16.6	313.9 ± 21.5	405.7 ± 32.3

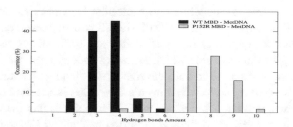

Fig. 2. Histogram of the hydrogen bond occurrence between methylated DNA and wild-type or P152R MBD (calculated over a period of 50 ns).

The histogram indicates that for most of the MD simulation period, the quantity of hydrogen bonds between the methylated DNA and wild-type MBD tends to be about 3–4 per sample, instead of 6–8 for the mutated MBD.

3.3 Binding Energy

The curves in Fig. 3 confirm the difference between Gibbs Free Energies (ΔGs) from the initial separated state to the final bound state for WT MBD binding to the methylated DNA and to the non-methylated DNA is approximately 15 kcal/mol. Furthermore, the results also show that the DNA-binding affinity of the mutant MBD is higher than that of the wild-type. The ΔGs for the methylated DNA binding are about 55 kcal/mol for the mutant, and about 40 kcal/mol for the wild-type. Meanwhile for the non-methylated DNA, the ΔG for the mutant (44 kcal/mol) is about double that of the wild-type (22 kcal/mol). These results indicate that the presence of the P152R mutation within the MBD causes the protein to be more adhesive to the DNA. Furthermore, for

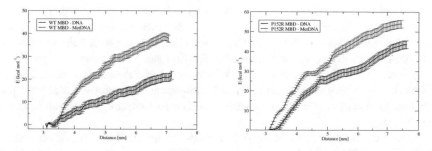

Fig. 3. Potential of Mean Forces (PMF) for the unbiased complexation.

the P152R MBD, the mutant domain keeps its binding preference for methylated DNA, but the ΔG is only about 20% higher than that for binding to the non-methylated DNA, instead of 70% for the wild-type.

4 Conclusions

According to these original modeling and Molecular Dynamics studies, the introduction of the P152R mutation into the MBD of MeCP2 protein causes moderate structural modifications, but not only localized within its vicinity. It affects the whole protein surface, including its opposite side which makes contact with the DNA. The mutation appears to result in an increase in both the size of the binding area and the strength of actual binding. Moreover, selectivity towards DNA that is methylated appears to be reduced. In essence, these simulations support the hypothesis that the P152R mutation will increase adhesiveness to methylated and non-methylated DNA and decrease the selectivity of MeCP2 to methylated DNA. This may have considerable biological implications for RTT. As MeCP2 is a gene expression regulator. Unbalanced binding affinities and lose of selectivity can lead to out of control gene expression - which is a scenario consistent with RTT symptoms.

If this prediction about adhesiveness and less selectiveness can be confirmed in further experimental assays, the consequences of the presence of this mutation seem quite clear: improper DNA-protein bindings, unbalanced regulation of gene expression, and subsequently RTT symptoms.

No previous study has further investigated the role of the P152R mutation in terms of effects on DNA binding. While the methodology used in this study only allows one to generate hypotheses, the results point towards a potential influence of this mutation on DNA binding and specifically that it may increase the adhesiveness and decrease the selectiveness of this protein. This hypothesis cannot ignore the fact that the P152R mutation affects other chromatin protein interactions, but it could explain experimental results that show that P152R MeCP2 has increased ability for binding to chromocenters, but impaired ability in terms of clustering them [16, 17].

When considered alongside previous studies, the results of this study might therefore form the basis of a set of testable hypotheses for future experimental studies relating to RTT. Such experiments might include an investigation into the role of other missense mutations, and experiments investigating the whole protein to check whether other components (particularly TRD presence) affect outcomes. Furthermore, high throughput screening or virtual screening, although expensive in the former case, may prove valuable in uncovering a ligand that might recover proper protein functionality. This could eventually be an important consideration since repository therapy alone is not an option in RTT due to challenges relating to potential over-dosage and specific delivery issues. The use of small chemical compounds for restoring wild-type properties and functionality to mutant proteins is becoming a common approach in rational drug design.

In terms of limitations, this study only considered the MBD of MeCP2 and this was mainly to be consistent with the original X-ray complex structure obtained by Ho et ali. [9]. There remains a risk that outcomes could change if other components or interactions of the protein (e.g. the TRD) are considered. The same can also be said of the natural 6xHis sequence at the protein C-terminal. It is expected there would be some extra attraction since the histidines are positive and DNA is in an overall sense negatively charged.

A further limitation is that this study is based on modeling and so, by its very nature can only ever result in predictions. In defense of this point, the study considers entropy effects and uses PMF to avoid problems associated with large free energy that is associated with standard decoupling schemes [18]. Because of these limitations one of the most important steps that needs to be taken is confirmation of these predictions and hypotheses in in vitro studies.

This is the first time that the P152R mutation and consequent protein binding dynamics have been explored this way. The insight that an MBD-MeCP2 mutation may cause an enhancement to the DNA binding affinity and also a decrease in selectivity to the methylated DNA could be important as it might provide a target for potential modification and rational drug design.

Acknowledgments. This work was in part supported by Fundação de Amparo à Pesquisa do Estado de Minas Gerais (FAPEMIG).

References

1. Guy, J., Gan, J., Selfridge, J., et al.: Reversal of neurological defects in a mouse model of Rett syndrome. Science **315**, 1143–1147 (2007). https://doi.org/10.1126/science.1138389
2. Neul, J.L., Fang, P., Barrish, J., et al.: Specific mutations in methyl-CpG-binding protein 2 confer different severity in Rett syndrome. Neurology **70**, 1313–1321 (2008). https://doi.org/10.1212/01.wnl.0000291011.54508.aa
3. Christodoulou, J., Grimm, A., Maher, T., Bennetts, B.: RettBASE: the IRSA MECP2 variation database-a new mutation database in evolution. Hum. Mutat. **21**, 466–472 (2003). https://doi.org/10.1002/humu.10194
4. Amir, R.E., Van den Veyver, I.B., Wan, M., et al.: Rett syndrome is caused by mutations in X-linked MECP2, encoding methyl-CpG-binding protein 2. Nat. Genet. **23**, 185–188 (1999). https://doi.org/10.1038/13810
5. Kudo, S., Nomura, Y., Segawa, M., et al.: Heterogeneity in residual function of MeCP2 carrying missense mutations in the methyl CpG binding domain. J. Med. Genet. **40**, 487–493 (2003)
6. Ballestar, E., Yusufzai, T.M., Wolffe, A.P.: Effects of Rett syndrome mutations of the methyl-CpG binding domain of the transcriptional repressor MeCP2 on selectivity for association with methylated DNA. Biochemistry **39**, 7100–7106 (2000)
7. Lemkul, J.A., Bevan, D.R.: Assessing the stability of Alzheimer's amyloid protofibrils using molecular dynamics. J. Phys. Chem. B **114**, 1652–1660 (2010). https://doi.org/10.1021/jp9110794
8. Hub, J.S., de Groot, B.L., van der Spoel, D.: g_wham—a free weighted histogram analysis implementation including robust error and autocorrelation estimates. J. Chem. Theory Comput. **6**, 3713–3720 (2010). https://doi.org/10.1021/ct100494z
9. Ho, K.L., McNae, I.W., Schmiedeberg, L., et al.: MeCP2 binding to DNA depends upon hydration at methyl-CpG. Mol. Cell **29**, 525–531 (2008). https://doi.org/10.1016/j.molcel.2007.12.028

10. Guex, N., Peitsch, M.C.: SWISS-MODEL and the Swiss-Pdb Viewer: an environment for comparative protein modeling. Electrophoresis **18**, 2714–2723 (1997). https://doi.org/10.1002/elps.1150181505

11. Case, D.A., Cheatham, T.E., Darden, T., et al.: The Amber biomolecular simulation programs. J. Comput. Chem. **26**, 1668–1688 (2005). https://doi.org/10.1002/jcc.20290

12. DeLano, W.L.: The PyMOL Molecular Graphics System, Version 1.1. Schrödinger LLC (2002). http://www.pymol.org

13. Duan, Y., Wu, C., Chowdhury, S., et al.: A point-charge force field for molecular mechanics simulations of proteins based on condensed-phase quantum mechanical calculations. J. Comput. Chem. **24**, 1999–2012 (2003). https://doi.org/10.1002/jcc.10349

14. Hess, B., Kutzner, C., van der Spoel, D., Lindahl, E.: GROMACS 4: algorithms for highly efficient, load-balanced, and scalable molecular simulation. J. Chem. Theory Comput. **4**, 435–447 (2008). https://doi.org/10.1021/ct700301q

15. Marquardt, D.W.: An algorithm for least-squares estimation of nonlinear parameters. J. Soc. Ind. Appl. Math. **11**, 431–441 (1963). https://doi.org/10.1137/0111030

16. Klose, R.J., Sarraf, S.A., Schmiedeberg, L., et al.: DNA binding selectivity of MeCP2 due to a requirement for A/T sequences adjacent to methyl-CpG. Mol. Cell **19**, 667–678 (2005). https://doi.org/10.1016/j.molcel.2005.07.021

17. Agarwal, N., Becker, A., Jost, K.L., et al.: MeCP2 Rett mutations affect large scale chromatin organization. Hum. Mol. Genet. **20**, 4187–4195 (2011). https://doi.org/10.1093/hmg/ddr346

18. Hoover, W.: Canonical dynamics: Equilibrium phase-space distributions. Phys. Rev. A **31**, 1695–1697 (1985). https://doi.org/10.1103/PhysRevA.31.1695

Impact of Genealogical Features in Transthyretin Familial Amyloid Polyneuropathy Age of Onset Prediction

Maria Pedroto[1,2]([⊠]), Alípio Jorge[1], João Mendes-Moreira[1], and Teresa Coelho[2]

[1] 1-LIAAD/INESC TEC, University of Porto, Porto, Portugal
maria.j.pedroto@inesctec.pt, amjorge@fc.up.pt, jmoreira@fe.up.pt
[2] Unidade Corino de Andrade (UCA), Centro Hospitalar do Porto (CHP), Largo
Prof. Abel Salazar, 4099-001 Porto, Portugal
https://www.inesctec.pt

Abstract. Transthyretin Familial Amyloid Polyneuropathy (TTR-FAP) is a neurological genetic disease that propagates from one family generation to the next. The disease can have severe effects on the life of patients after the first symptoms (onset) appear. Accurate prediction of the age of onset for these patients can help the management of the impact. This is, however, a challenging problem since both familial and non-familial characteristics may or may not affect the age of onset. In this work, we assess the importance of sets of genealogical features used for Predicting the Age of Onset of TTR FAP Patients. We study three sets of features engineered from clinical and genealogical data records obtained from Portuguese patients. These feature sets, referred to as *Patient, First Level* and *Extended Level* Features, represent sets of characteristics related to each patient's attributes and their familial relations. They were compiled by a Medical Research Center working with TTR-FAP patients. Our results show the importance of genealogical data when clinical records have no information related with the ancestor of the patient, namely its Gender and Age of Onset. This is suggested by the improvement of the estimated predictive error results after combining First and Extended Level with the Patients Features.

Keywords: Genealogical data · Regression algorithms
Relevancy estimation · Feature construction

1 Introduction

TTR-FAP is a neurological genetic disease [1] propagated from parents to offspring that is currently present in clusters of patients in Portugal, Sweden, France and Brazil [2]. The first patient was diagnosed in 1952 by Dr. Mário Corino de Andrade in Hospital Santo António, and the first documented clinical case was subsequently presented in [3].

© Springer Nature Switzerland AG 2019
F. Fdez-Riverola et al. (Eds.): PACBB 2018, AISC 803, pp. 35–42, 2019.
https://doi.org/10.1007/978-3-319-98702-6_5

In genealogical studies of hereditary diseases, an important aspect of study is the estimation and evaluation of a Patient's Symptom Onset, which is the age in which a patient first experiences the symptoms of a disease. When reaching this landmark, different sets of medical procedures are triggered for a more thorough follow-up procedure.

For Portuguese TTR-FAP patients, the most extensive work related with Analyzing the Age of Onset Variability was performed by [4] and the results are focused on the statistical comparison of Portuguese and Swedish Patients characteristics.

In this work, we estimate the predictive importance of three sets of features used in the regression approach to predict the Age of Onset of Patients diagnosed with TTR-FAP. We test the following hypotheses: (H1) whether extending the extension of the basic set of patient features with first level genealogical features improves the results of the best performing regression algorithms; and (H2) whether the further extension of the set of features with extended genealogical features improves the results of the regression algorithms. We have organized this work as follows: in Sect. 1 we introduce our work and instantiate our field of work; in Sect. 2 we give a general overview of the predictive methods considered; in Sect. 3 we define our sets of features, modeling scenario, predictive methodology, estimation and evaluation approach; in Sect. 4 we present our results and discuss their values; while in Sect. 5 we assess our main conclusions and future work directions.

2   Regression Algorithms

In this section we describe relevant Regression algorithms, separating them between state of the art Machine Learning Algorithms and specific feature selection approaches [5]. The chosen algorithms are: Elastic Net [6], Lasso (L1-regularization, also known as Least Absolute Shrinkage and Selection Operator) [7], Ridge (L2-regularization) [8], Linear Regression, Decision Tree, Random Forest, and Support Vector Machines. For a complete study on Regression methods and their applications we suggest [9].

2.1   Regression Algorithms (State of the Art Approaches)

Regression Algorithms attempt to explicitly model the relationship between inputs or independent variables and the outputs or dependent target variables, typically in the form of parametric equations. These equation parameters are estimated from the data [10].

Generally Linear Regression predicts a single target variable based on one or more attributes modeled by a linear relationship. The regression coefficients can be inferred with the least squares method to minimize the errors [11].

Using binary recursive partitioning, the purpose of Decision Tree Regressors is to predict numerical target variables by dividing the predictor variables in non-overlapping regions. The partitioning continues until all the records in each leaf-node have the same target value or until the size limit is reached. If the later happens, the predicted value of a new record is calculated by averaging the target values of all the records that share the same leaf.

Random Forests generally improve the performance achieved by different sets of Decision Trees, thus capturing more intrinsic properties in a dataset. Generally a number of Decision Trees are generated and trained using each a different bootstrap sample of the training data and selecting randomly a subset of features at each split node in order to decide the split criterion. Typically, the number of predictive variables to be chosen at each split node is equal to the squared root of the total number of predictors. After the training phase, the prediction is achieved by applying a new record to all the trees and returning the average values achieved.

In the special case of SVR, and supposing that we have a set of different training instances, we define a limit (ϵ) with the purpose of finding a function f(x) that has at most ϵ deviation from the actually obtained targets (yi) for all the training data, and at the same time is as flat as possible. In other words, we do not care about errors as long as they are less than (ϵ), but will not accept any deviation larger than this [12].

2.2 Methods Which Incorporate Feature Selection

A common problem in statistical modeling is variable selection, that is, which predictive variables should be retained in a model. There are several reasons related with the importance of variable selection. A possible strategy for variable selection is to use regularized regression methods (also called penalized regression or shrinkage regression methods). These methods fit generalized linear models for which the sizes of the coefficients are constrained. But they come at a cost, as most search techniques demand a tuning approach capable of searching for the optimal subset of relevant features. This introduces additional complexity to the knowledge data discovery pipeline, as well as increasing the general computational cost [5]. In our case, mostly due to computational costs, our general experimental approach relies on comparing their bulk estimated predictive value when adding classes of features. Next we will define the shrinkage methods analyzed.

When dealing with regression problems it is important to consider what to do in case of: (i) over-fitting, that is when a model does not generalize well and can be too attached to specific conditions of the training set; (ii) collinearity, that is when a predictor variable can be predicted by using other predictive variables; (iii) large number of variables or low ratio of instances per number of variables; or (iv) to simplify the model explanation, as some types of algorithms can generate solutions that are too complex when we have a high number of predictive variables. To deal with a few of these problems one can apply Shrinkage methods [9]. In our case we used Elastic Net, Lasso and Ridge algorithms.

In the case of *Lasso Regression* (L1-regularization), it helps to perform feature selection in sparse feature spaces. It is useful in some contexts, as it reduces the number of variables upon which the given solution is dependent. Thus, it is good for variable selection and bad for group selection when there are predictive variables strongly correlated. *Ridge regression* (L2-regularization) addresses some of the problems of Linear Regression by imposing a penalty on the size of coefficients. It is good for multicollinearity, and group feature selection, while being not so good for variable selection. *Elastic Net* is a model that is capable of dealing with collinearity issues that need variable selection (combines the strengths of Lasso and Ridge algorithms) [6].

3   Methodology and Experimental Context

In this section we describe the methodology to evaluate the predictive importance of three sets of genealogical features, namely Patients, First Level and Extended Level sets (see Table 1). Since there are many missing values in our data, along with other data quality problems, in this work we were interested in evaluating how predictions changed when adding specific sets of features. This way we have three sets of experiments: Dataset A, which is a dataset containing patients with previously diagnosed ancestor; Dataset B, which contains records of Patients for which we do not have a measurable Age of Onset for the Ancestor; and Full Dataset, which contains records from Dataset A and Dataset B. In Dataset A we have 1103 patients while in Dataset B we have 1709. To deal with missing data we used: (i) the Missing-Indicator method [13], that is for each column with missing values we would generate a binary auxiliary column with 1 when there is information and 0 when there is no available information, while each missing value would be encoded with −100 value. To perform our experiments, we chose to use records of Asymptomatic Patients when they were 20 and 30 years old, as these represent important social thresholds.

3.1   Approach and Regression Algorithms Used

Taking our sets of features into account, we built virtual scenarios for a set of different ages of Patients. This allowed us to simulate age of onset prediction cases for patients, along with their lifespan. This means that, for each of our asymptomatic patients, we took into consideration the known data for the years in which they were 20 and 30 years-old (if still asymptomatic), and instantiated all of our different sets of features. This way, prediction at each age can be seen as a specific and individual regression predictive task.

3.2   Modeling Schema and Experiments Setup

To estimate the predictive value of each set of features, we created a pipeline for model tuning, selection and accuracy estimation. Afterwards, we ran the different sets of experiments through the pipeline and compared the results using the Mean Absolute Error (MAE) measure. Our model selection schema is defined in a top-level procedure that, for a given numerical age, generates an instance of a model. We integrated two k-fold Cross Validation procedures into what

Table 1. List of engineered features

Type	Description	N. Feat.
Patient	Patients' Gender	1
	Father and Mother Age Onset	1
	Born before Father or Mother had Symptoms	1
	Age of Onset (Target Variable)	1
First Level	Number of Children (by gender)	2
	Number of Siblings (by gender)	2
	Number of Uncles and Aunts	2
	Number of Children with the disease (by gender)	2
	Number of Siblings with the disease (by gender)	2
	Number of Uncles and Aunts with the disease	2
	Avg, Max and Min Age Onset of Children (by gender)	6
	Avg, Max and Min Age Onset of Siblings (by gender)	6
	Avg, Max and Min Age Onset of Uncles and Aunts	6
Extended	Avg, Max and Min Age Onset of Patients followed at the Clinical Unit (by gender)	6
	Avg, Max and Min Age of Onset of Patients in the Family Tree (by gender)	6
	Avg, Max and Min Age of Onset of Early Onset Patients (Patients with Age of Onset < 50)	6
	Avg, Max and Min Age of Onset Late Onset Patients (Patients with Age of Onset ≥ 50)	6

is known as Nested Cross Validation. For reference purposes, a single K-Fold Cross Validation consists of dividing the set of examples into K subsets with approximately the same size. Then, a prediction algorithm takes K-1 partitions to train and estimates the accuracy on the last fold. The process is repeated K times, using a different partition to measure the accuracy each time. The final evaluation results consider the average of the errors observed in each of the K test sets [9]. In the case of Nested Cross Validation, and for each training phase, we have another cross validation to find the set of best tuning parameters for a specific algorithm. This way we are able to perform a model selection between different algorithms, and tune their parameters towards a specific domain. The full algorithm is presented in [14]. After both the tuning of the parameters and the model selection are completed, the predictive accuracy for each model is estimated by the MAE calculated in the outer folds. In the case of our experimental setup, for each individual age group (20 and 30 years-old datasets), and for each individual dataset (Dataset A, Dataset B, Full Dataset), we performed a total of 4 experiments. These were: (i) (Patient Features) hereafter designated as Experiment 1; (ii) (Patient + First Level Features) hereafter designated as Experiment 2; (iii) (Patient + Extended Features) hereafter designated as Experiment 3; (iv) (Patient + First Level + Extended Features) hereafter designated as Experiment 4.

4   Results and Discussion

The results for both the ages of 20 and 30 experiments are expressed in Fig. 1. An analysis of the individual boxplots show us a clear improvement of the predictive accuracy of Dataset B and Full Dataset when we increase the number of features. To check its statistical significance, we used a Wilcoxon Hypothesis Rank test and performed 4 individual comparisons, namely: (i) (A): Experiment 1 versus Experiment 2; (ii) (B): Experiment 1 versus Experiment 3; (iii) (C): Experiment 2 versus Experiment 4; (iv) (D) Experiment 3 versus Experiment 4.

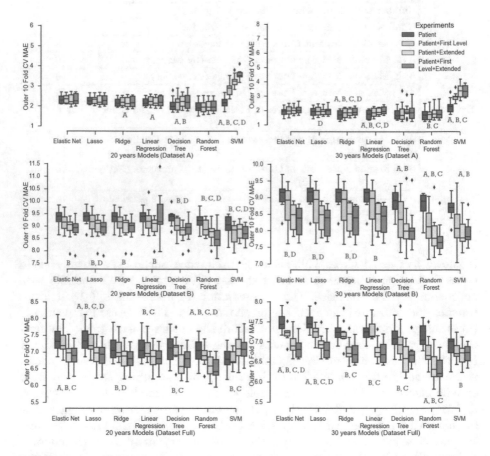

Fig. 1. 20 years (left) and 30 years (right) experiments, for Dataset A, Dataset B and Full Dataset from top to bottom. For each algorithm there are a total of 4 experiments. When there's a letter (A, B, C, D) near to the boxplot this means that there is a significant difference in the set of results between: (A): Experiment 1 versus Experiment 2; (B): Experiment 1 versus Experiment 3; (C): Experiment 2 versus Experiment 4; and (D) Experiment 3 versus Experiment 4.

Wilcoxon Hypothesis Rank test is a non-parametric statistical hypothesis test used to compare if two related samples, or repeated measurements on a single sample mean, rank values differ. These results, when there are significant statistical differences, are represented in the graphs next to each individual algorithm as (A), (B), (C) and (D). When we do not have a letter in the graph it means that we reject the null hypothesis. On the other hand if we have a letter, it means that we are confident, with a 95% confidence interval, of the existence of significant differences in the rank. In this case the main observations are: (i) (Dataset A): when we integrate extra information to the Patients Features there is a general loss of accuracy for the SVR algorithm; (ii) (Dataset A): the error amplitude is smaller for shrinkage methods, which leads us to conclude that these methods are important for cases with high number of features that have possible correlations between them; (iii) (Dataset B and Full Dataset): there is an increase in the amplitude of the errors for most algorithms when results are compared with Dataset A; and (iv) (Dataset B and Full Dataset): there is a general increase in the number of outliers present, which leads us to conclude that there is a significant difference in the distribution of the training and testing datasets for each fold.

With these sets of results and regarding our previously defined research hypothesis (H1) and (H2) we conclude that: (i) in the dataset A, results tend to get worse when we add First Level and Extended Level Features (this is understandable as this dataset is expected to have the more relevant characteristics in the Patients Feature Set); (ii) in the case of Dataset B and Full Dataset, we have an evident improvement in the predictive accuracy by adding First Level features; and (iii) there are high differences in the reaction of the algorithms to each set of features.

5   Conclusions and Future Work

In this work we evaluated and compared the predictive accuracy of three sets of features used to Predict the Age of Onset of TTR-FAP patients. We defined a set of experiments in order to determine if, on average, the rank results of a set of Nested Cross Validation experiments change. The fact that there are large differences in the results of the algorithms in each set of experiments suggests that we need to perform a thorough test to determine the behavior of the algorithms to each feature. This is more important in the case of the shrinkage algorithms, as these can give insights on the performance of the considered sets of features. Regarding our research hypothesis, we verified that the addition of first level genealogical features does indeed improve the predictive accuracy of the tested Regression Algorithms for the Dataset B and for the Full Dataset (H1). We also verified that the further extension of features with extended genealogical characteristics also improves the results in both Dataset B and Full Dataset (H2). Further developments can be directed into: (i) studying the impact of the features in each algorithms; and (ii) improving the Dataset A results with the First Level and Extended Features sets.

Acknowledgements. This work is financed by the ERDF - European Regional Development Fund through the Operational Programme for Competitiveness and Internationalisation - COMPETE 2020 Programme within project POCI-01-0145-FEDER-006961, by National Funds through the FCT - Fundação para a Ciência e a Tecnologia (Portuguese Foundation for Science and Technology) as part of project UID/EEA/50014/2013 and by Centro Hospitalar do Porto (ChPorto) through grant BI.09/2015/UCA/CHP.

References

1. Ando, Y., Coelho, T., Berk, J.L., Cruz, M.W., Ericzon, B.-G., Ikeda, S., Lewis, W.D., Obici, L., Planté-Bordeneuve, V., Rapezzi, C., Said, G., Salvi, F.: Guideline of transthyretin-related hereditary amyloidosis for clinicians. Orphanet J. Rare Dis. **8**, 31 (2013)
2. Parman, Y., Adams, D., Obici, L., Galán, L., Guergueltcheva, V., Suhr, O.B., Coelho, T.: Sixty years of transthyretin familial amyloid polyneuropathy (TTR-FAP) in Europe. Curr. Opin. Neurol. **29**, S3–S13 (2016)
3. Andrade, C.: A peculiar form of peripheral neuropathy; familiar atypical generalized amyloidosis with special involvement of the peripheral nerves. Brain J. Neurol. **75**(3), 408–427 (1952)
4. Sousa, A.M.B.C.: A Variabilidade Fenotípica da Polineuropatia Amiloidótica Familiar: um estudo de genética quantitativa em Portugal e na Suécia. Ph.D. thesis, University of Porto (1995)
5. Saeys, Y., Inza, I., Larrañaga, P.: A review of feature selection techniques in bioinformatics. Bioinformatics **23**, 2507–2517 (2007)
6. Zou, H., Hastie, T.: Regularization and variable selection via the elastic net. J. R. Stat. Soc. B **67**(2), 301–320 (2005)
7. Tibshirani, R.: Regression shrinkage and selection via the Lasso. J. R. Stat. Soc. **58**, 267–288 (1996)
8. Hoerl, A.E., Kennard, R.W.: Ridge regression: biased estimation for nonorthogonal problems. Technometrics **12**, 55–67 (1970)
9. Hastie, T., Tibshirani, R., Friedman, J.: The Elements of Statistical Learning: Data Mining, Inference, and Prediction. Springer Series in Statistics, 2nd edn. Springer, New York (2009)
10. Dasgupta, A., Sun, Y.V., König, I.R., Bailey-Wilson, J.E., Malley, J.D.: Brief review of regression-based and machine learning methods in genetic epidemiology: The Genetic Analysis Workshop 17 experience. Genet. Epidemiol. **35**, S5–S11 (2011)
11. Schneider, A., Hommel, G., Blettner, M.: Linear regression analysis. Dtsch Arztebl Int. **107**(44), 776–782 (2010)
12. Smola, A.J., Schölkopf, B.: A tutorial on support vector regression *. Stat. Comput. **14**(3), 199–222 (2004)
13. Groenwold, R.H.H., White, I.R., Donders, A.R.T., Carpenter, J.R., Altman, D.G., Moons, K.G.M.: Missing covariate data in clinical research: When and when not to use the missing-indicator method for analysis. CMAJ **184**, 1265–1269 (2012)
14. Petersohn, C.: Temporal Video Segmentation. Jörg Vogt Verlag, Dresden (2010)

Feature Selection and Polydispersity Characterization for QSPR Modelling: Predicting a Tensile Property

Fiorella Cravero[1,2], Santiago Schustik[1,2], María Jimena Martínez[3],
Carlos D. Barranco[4], Mónica F. Díaz[1,5], and Ignacio Ponzoni[3(✉)]

[1] Planta Piloto de Ingeniería Química – PLAPIQUI (UNS-CONICET),
Bahía Blanca, Argentina
[2] Comisión de Investigaciones Científicas (CIC), Bahía Blanca, Argentina
[3] Instituto de Ciencias e Ingeniería de la Computación (UNS–CONICET),
Departamento de Ciencias e Ingeniería de la Computación,
Universidad Nacional del Sur (UNS), Bahía Blanca, Argentina
ip@cs.uns.edu.ar
[4] Intelligent Data Analysis (DATAi), Division of Computer Science,
Pablo de Olavide University, 41013 Seville, Spain
[5] Departamento de Ingeniería Química, UNS, Bahía Blanca, Argentina

Abstract. QSPR (Quantitative Structure-Property Relationship) models proposed in Polymer Informatics typically use reduced computational representations of polymers for avoiding the complex issues related with the polydispersion of these industrial materials. In this work, the aim is to assess the effect of this oversimplification in the modelling decisions and to analyze strategies for addressing alternative characterizations of the materials that capture, at least partially, the polydispersion phenomenon. In particular, a cheminformatic study for estimating a tensile property of polymers is presented here. Four different computational representations are analyzed in combination with several machine learning approaches for selecting the most relevant molecular descriptors associated with the target property and for learning the corresponding QSPR models. The obtained results give insight about the limitations of using oversimplified representations of polymers and contribute with alternative strategies for achieving more realistic models.

Keywords: QSPR · Polydispersity · Artificial intelligence · Feature selection

1 Introduction

The development of machine learning tools for cheminformatics is a dynamic area of research. In particular, the use of artificial intelligence approaches for inferring QSPR (Quantitative Structure-Property Relationship) models has been increasing considerably through last decade [1]. The design of QSPR models constitutes a particular case of predictive modeling problem, where a domain expert is focused on discovering the relationship between some molecular descriptors and a target variable. Molecular descriptors are variables that characterize the structure of the chemical compounds, and

© Springer Nature Switzerland AG 2019
F. Fdez-Riverola et al. (Eds.): PACBB 2018, AISC 803, pp. 43–51, 2019.
https://doi.org/10.1007/978-3-319-98702-6_6

the first step for inferring a QSPR model is to identify which descriptors are more related with the property under study. Software tools for molecular descriptors computation can calculate thousands of variables but, in general, a regression QSPR model only requires a short number of descriptors for estimating a target property. Therefore, the selection of descriptors in QSPR modeling is an instance of the traditional feature selection (FS) problem.

A particular complex scenario of descriptors selection for QSPR modelling occurs in Polymer Informatics. This interdisciplinary science requires a judicious computational modelling of synthetic polymers, which are chain-like molecules consisting of one or more structural repeat units (SRUs) [2]. A polymeric material is made of several polymeric chains with different lengths and molecular weights. Therefore, in contrast with a typical drug molecule, a polymeric material is better characterized by a distribution of molecular weights instead of a single molecular weight value. This distinctive attribute of the polymeric materials is known as polydispersion and, as a consequence of this aspect, each molecular descriptor of a polymeric material has also associated a discrete distribution of values that it is obtained by calculating the molecular descriptor for the different polymeric chains and its frequencies. Nevertheless, traditional QSPR approaches proposed for predicting polymer properties in Material Science in general do not take into account the polydispersion, oversimplifying the computational representation of each polymeric material to its SRU [3, 4]. In other words, these QSPR models are inferred from datasets where the descriptors are only computed for the shorter polymeric chain that characterize each material, without considering the remaining polymeric chain lengths neither its associated frequencies.

In this work, our main goal is to analyze the impact of this oversimplification in the computational representation of polymeric materials and to evaluate alternative strategies for addressing the molecular descriptors selection problem in a context of polydispersion. In the next section, the proposed methodology for the analysis is explained. After that, preliminary results for the prediction of a material property known as *elongation at break*, using an in-house polymeric materials dataset, are presented. Finally, conclusions and potential future directions are discussed.

2   Feature Selection in QSPR for Polymeric Informatics

The techniques based on QSPR principles estimate a property from molecular, structural and nonstructural descriptors that numerically quantify different issues of a molecule [5]. In mathematical terms, a QSPR model is defined as a function $Y = f(X)$, where $X = (x_1, x_2, \ldots, x_n)$ is a chemical compound database represented as a vector of molecular descriptors and Y is an experimental target property. The aim is to infer f from a dataset with chemical compounds, where a number of molecular descriptors are computed for each compound using specific tools like DRAGON [6]. Besides, experimental data is also required for the physicochemical property or biological activity of interest (Y). From this dataset the function f can be learned by using a training method. Once f has been inferred, this function may be applied to new compounds not covered by training. Thus, f can predict *in silico* the value of a property based on the analysis of data from other experiments. To assess this function f is necessary to identify, first, which molecular descriptors are related to the property under study.

In the case of Polymeric Materials, these macromolecules are created via polymerization of many structural repeated units (SRUs), also known as monomers. The polymerization produces chains with different lengths. For this reason, the polymers are polydispersed. In other words, they have associated more than one molecular weight as it was mention before. In Fig. 1(a), a typical curve of molecular weight dispersion is presented. On the x-axis, the molecular weight is represented, and on the y-axis represents the frequency of occurrence of the chains with that length. The average weights are used to describe a polymer, usually the Mw (weight-average molar mass) and the Mn (number-average molar mass). In Fig. 1(b), a step of the polymerization procedure is illustrated. This software uses the Simplified Molecular Input Line Entry Specification (SMILES) notation, which represents molecules as character strings [7, 8], for building polymer chains of different lengths executing multiple sequential concatenations of the SRU by head and tail as it will be explained in Sect. 3.

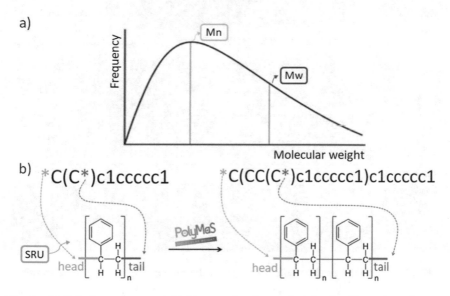

Fig. 1. (a) Polydispersity curve. (b) Computational procedure for *in silico* polymerization.

3 Proposed Methodology and Experimental Results

In Polymer Informatics, obtaining material databases is a complex task. Moreover, if the database must be integrated by polymers related to the study of mechanical properties associated to a tensile test. In this work, an in house database developed by our research group has been used [9]. The polymers in this database are homopolymers, linear and amorphous. The 77 initial polymers were characterized by their SRU in SMILES code. To get the databases used in this work, it was necessary to polymerizate molecules that reach the average weights of the polymers in the database. The Mn varies in a range of 4700 to 765000 [g/mol] and the Mw does it with 19500-2200000 [g/mol]. The target property modelled in this work is related to the tensile test. In this test, a polymeric sample is subjected to a controlled strain (constant Cross Head Speed) until failure. Many mechanical properties are measured during the execution of this test. In particular, the elongation at break corresponds to the elongation of a material until the fracture occurs (breaking point) during the elongation of the sample (see Fig. 2 in the upper right corner).

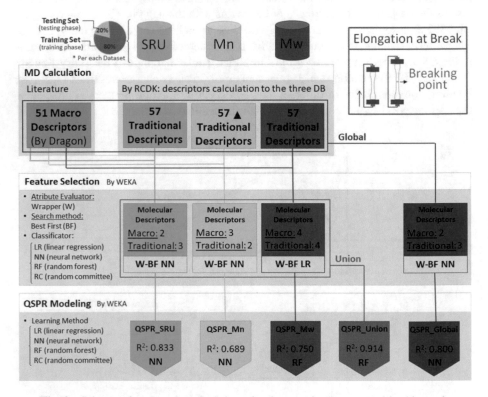

Fig. 2. Scheme of proposed methodology for the experiments reported in this work.

The proposed experimentation is schematized in Fig. 2. The initial Data Base (DB) consist of 77 polymers represented by their SRU, whose change length ranges are [5–7205 SRU] to get Mn and [24–10773 SRU] to get Mw. The SMILES representation allows retain indicators that locate the extremes (Head and Tail) from the SRU. Our group developed a Polymer Maker Smiles-based (PolyMAS) software that performs a Head-Tail polymerization using SMILES representation from each SRU. PolyMAS joins the Head of one SRU with the Tail from another SRU, repeating this process until obtaining the desired chain weight (see Fig. 1b). Hence, PolyMAS allows to obtain the SMILES codes corresponding to the weights Mn and Mw for each polymer in the DB. Then each one of these codes, together with the SMILES codes from the SRU, are the input for the Molecular Descriptor calculator algorithm (using RCDK library in R [10]).

This way the molecular descriptors was obtained for three considered weights (SRU, Mn and Mw), which resulted in three different DB.

Although the initial polymers quantity was 77 and the maximum number of descriptors calculated was 302, due to limitations of the molecular descriptors calculation algorithms, for some polymers it was not possible to calculate any descriptor and for other polymers only a reduced set of descriptors was calculated, therefore polymers and descriptors were filtered, reducing each DB to 61 polymers and 57 descriptors. To complete each DB, 51 macro molecular descriptors were added. These include molecular descriptors with a macro view, parameter values of the tensile test and different average weights [3]. After analyzing the structure of each polymer in the DB, four polymers were detected as outliers, because they are considerably different from the rest of the DB in terms of their molecular structures. These outliers were eliminated. Therefore, the three final databases (SRU, Mn and Mw) are integrated by 57 polymers and 108 descriptors. Additionally, a global database was defined by including all molecular descriptors computed for the SRU, Mn and Mw databases. Therefore, this fourth database contains the information associated to the three different instances of molecular weight of polymeric materials. This last database constitutes a first try for characterize the polymeric materials by capturing part of their polydispersion. Each database has been divided into two parts, one part $\sim 80\%$ (46 molecules) for the training phase (selection of descriptors and model making) and the other, $\sim 20\%$ (11 molecules), for testing.

WEKA tool [11] was used for selecting the most relevant molecular descriptors from each database. These feature selection experiments had been carried out using a wrapper method, using Best First as search algorithm (W-BF) and four different classification techniques; Linear Regression (LR), Neural Networks (NN), Random Forest (RF) and Random Committee (RC). Once the four subsets of molecular descriptors had been selected for each dataset, the domain expert chose the best subsets taking into account cardinality and balance between the different classes of descriptors. In other words, the QSPR modeler's goal consists of obtaining a subset with few molecular descriptors which belong to traditional and macro descriptor classes in similar proportions.

Table 1 shows the final subset of molecular descriptors chosen for each dataset. For the Global database, subscripts are used for indicating which weight instance version of the molecular descriptor has been selected. Descriptors selected for more than one dataset are highlighted in bold. In particular, five descriptors are selected more than once, including all descriptors selected from the Global database. This result reveals that some descriptors values are relevant for capturing how polymer chains of some specific length contributes to the estimation of the tensile strength at break, whereas other descriptors are relevant independently of the weight instance represented.

After the selection of the best molecular descriptor subsets, QSPR models were inferred from each database using the same four regression methods mentioned before (LR, NN, RF and RC). For the case of the Global database, two alternatives have been explored: to use the subsets of molecular descriptors selected from this database, and to use the union of the subsets of molecular descriptors selected from the other three databases (SRU, Mn and Mw). The correlation coefficient (R^2 value) of the best QSPR model inferred in each case, together with the regression method that achieves this

Table 1. Subsets of molecular descriptors selected for each database. The molecular descriptors shared by two or more subsets are highlighted in bold.

DB	Traditional	Macro	Cardinality
SRU	nRings4, **C2SP2**, **ALogP**	**nSASC**, **nRMC.nRSC**	5
Mn	nRings5, khs.dssC	nLog_PMC, **nSASC**, **nRMC.nRSC**	5
Mw	VAdjMat, nAtomP, khs.dsCH, **khs.aaO**	AMR, Log_PMC.Log_PSC, **nRMC. nRSC**, nPMC.nPSC	8
Global	**C2SP2.$_{SRU}$**, **ALogP.$_{SRU}$**, **khs.aaO.$_{Mw}$**	**nSASC.$_{Mw}$**, **nRMC.nRSC.$_{Mw}$**	5

Table 2. QSPR models with higher performances for each database. The subset Union represents the union set of the three subsets obtained by applying feature selection over the databases SRU, Mn and Mw. The last columns contain different error metrics: Mean absolute error (MAE), Root mean squared error (RMSE), Relative absolute error (RAE) and Root relative squared error (RRSE).

QSPR Model	DB	FS Method	Training	R^2	MAE	RMSE	RAE	RRSE
QSPR_SRU	SRU	W-BF NN	NN	0.8330	1.05	1.21	35.83	34.64
QSPR_Mn	Mn	W-BF NN	NN	0.6899	1.37	1.70	46.73	48.91
QSPR_Mw	Mw	W-BF LR	RF	0.7509	1.05	1.34	35.92	38.52
QSPR_Union	Global	Union	RF	0.9147	0.29	0.39	47.64	51.27
QSPR_Global	Global	W-BF NN	NN	0.8008	1.46	1.63	49.92	46.75

accuracy, are denoted at the end of Fig. 2. Table 2 shows additional metrics of these models.

The next step of the proposed methodology consists of evaluating the performances of the QSPR models specific of each weight instance (QSPR_SRU, QSPR_Mn and QSPR_Mw), when they are applied using each of these databases. The main goal of this experiment is to assess the performance of a QSPR model generated for a specific chain length representation of the polymers database when the model is applied over a different characterization of the same polymers database. Figure 3 shows the correlation coefficients of this combinatorial experiment. From these results, it is clear that

Elongation at break	Models Performance (R^2)		
	QSPR_SRU	QSPR_Mn	QSPR_Mw
DataBase — SRU	0.833	0.7577	0.3137
DataBase — Mn	0.1217	0.6899	0.6528
DataBase — Mw	0.0314	0.4924	0.7509
Average (R^2)	0.3287	0.6466	0.5724

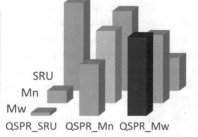

Fig. 3. Combinatorial experiments between QSPR models and databases. Performance metrics for testing datasets are reported as R^2 values.

when a QSPR model inferred for a given weight instance database is applied over a database that corresponds to a bigger weight instance representation, the model performance strongly falls. In contrast, when a QSPR model is applied over a database that corresponds to a lower weight instance representation, the model performance is reduced most softly and, even, in some cases can improve (for example, when $QSPR_{Mn}$ is applied over the SRU database). Therefore, we can conclude that the oversimplification of the polymer representations is an unadvisable practice for QSPR modelling, at least for this case of study.

From Table 2, analysing the results for the Global database, it is also possible to compare the performance of the QSPR_Global model learned by using the subset of five molecular descriptors specifically selected from this database (see descriptors in last row of Table 1) with the QSPR_Union model obtained by using the union of the subsets selected for databases SRU, Mn and Mw, which is integrated by fifteen molecular descriptors. As we can see, the QSPR_Union model clearly outperforms the QSPR_Global model, suggesting that combining the features learning for the different weight instance databases can be an effective strategy for dealing with polydispersion. Nevertheless, it is important to remark that this union model uses fifteen molecular descriptors in contrast with the only five variables included in the global model and, as it is well-known, large models can suffer overfitting.

Finally, the generalizabity of the QSPR models can be analyzed by contrasting the accuracies of QSPR_Union and QSPR_Global, reported in Table 2 as R^2, with the average accuracies of $QSPR_{SRU}$, $QSPR_{Mn}$, and $QSPR_{Mw}$, reported in Fig. 3. In this regard, it is important to note that QSPR_Union and QSPR_Global have been evaluated over the Global database and, therefore, in order to achieve a fair comparison in terms of generability, $QSPR_{SRU}$, $QSPR_{Mn}$, and $QSPR_{Mw}$ model accuracies obtained for the SRU, Mn and Mw databases must be averaged for assessing their global performances. The use of the average in this context is a sound decision, because the three databases have the same dimensionality (quantities of polymers and molecular descriptors) and, then, the same relative weight in the conformation of a global accuracy measure. From these values, it is clear that QSPR_Union and QSPR_Global outperform the precision of the QSPR models learned from polymer databases that correspond with only one weight instance, where the $QSPR_{SRU}$ has the worst performance (0.3287). Therefore, it can be concluded that models inferred from several weight instances have better generalizability properties.

4 Conclusions

In Polymer Informatics, an emerging subfield of Cheminformatics, the inference of QSPR (Quantitative Structure-Property Relationship) models constitutes a relevant topic associated with the design *in silico* of new industrial materials. A complex issue for computational modelling of synthetic polymers is related with the polydispersion that characterize these macromolecular structures. QSPR models proposed in the literature for predicting material properties avoid this issue, oversimplifying the computational representation of the macromolecules to structural repeat units (SRUs).

In this work, our contribution is to assess the effect of this simplistic vision of the polymer complexities and to propose new ideas for achieving other characterizations of the materials that capture, at least partially, their polydispersion. Specifically, a study for estimating a tensile property of polymers is presented, where four different computational representations are evaluated in combination with several machine learning methods for selecting the most relevant molecular descriptors associated with the target property and regression methods for inferring the QSPR models.

From the results, it is clear that oversimplification of the polymer representations is in general an unadvisable practice for QSPR modelling, at least under the scope of this case of study. Regarding alternative ideas for capturing polydispersion, we contribute with a database representation based in the calculation of the molecular descriptors for three polymeric chain lengths for each material (SRU, Mn and Mw), which achieves a high performance. In particular, it is clear that the QSPR models obtained from databases that included different weight instances of the polymers reach better generalizability properties. As future work, we plan to extend our in house database in order to improve the applicability domain of the QSAR models and also to evaluate the proposed representation strategy for other mechanical polymer properties.

Acknowledgments. This work is kindly supported by CONICET, grant PIP 112-2012-0100471 and UNS, grants PGI 24/N042 and PGI 24/ZM17. This work has been also partially supported by the Spanish Ministry of Economy and Competitiveness and the European Regional Development Fund under the project TIN2015-64776-C3-2-R DIFERENTIAL@UPO: Massive data management, filtering and exploratory analysis. We also thank to SGPEC of UNS for partially supported the visit of Dr. Barranco to the ICIC in 2016 and to the AUIP (*Asociación Universitaria Iberoamericana de Postgrado*) for partially supported the visit of Dr. Ponzoni to the Pablo de Olavide University in 2017.

References

1. Mitchell, J.B.O.: Machine learning methods in chemoinformatics. WIREs Comput. Mol. Sci. **4**, 468 (2014)
2. Audus, D.J., De Pablo, J.J.: Polymer informatics: opportunities and challenges. ACS Macro Lett. **6**, 1078–1082 (2017)
3. Palomba, D., Vazquez, G.E., Díaz, M.F.: Novel descriptors from main and side chains of high-molecular-weight polymers applied to prediction of glass transition temperatures. J. Mol. Graphics Model. **38**, 137–147 (2012)
4. Cravero, F., Martínez, M.J., Vazquez, G.E., Díaz, M.F., Ponzoni, I.: Feature learning applied to the estimation of tensile strength at break in polymeric material design. J. Integr. Bioinform. **13**, 286 (2016)
5. Le, T., Chandana, Epa V., Burden, F.R., Winkler, D.A.: Quantitative structure-property relationship modeling of diverse materials properties. Chem. Rev. **112**(5), 2889 (2012)
6. DRAGON, Version 5.5, Talete srl, Milan, Italy, (2007)
7. Weininger, D.: SMILES, a chemical language and information system. 1. Introduction to methodology and encoding rules. J. Chem. Inf. Comput. Sci. **28**, 31–36 (1988)
8. Weininger, D., Weininger, A., Weininger, J.L.: SMILES. 2. Algorithm for generation of unique SMILES notation. J. Chem. Inf. Comput. Sci. **29**, 97–101 (1989)

9. Palomba, D., Cravero, F., Vazquez, G.E., Diaz, M.F.: Prediction of tensile strength at break for linear polymers applied to new materials development. In: Proceeding of the International Congress of Metallurgy and Materials - Sam-Conamet, Santa Fe, Argentina (2014)
10. Guha, R.: Chemical Informatics Functionality in R. J. Stat. Software **6**(18), 1–16 (2007)
11. Hall, M., Frank, E., Holmes, G., Pfahringer, B., Reutemann, P., Witten, I.H.: The WEKA data mining software: an update. SIGKDD Explor. **11**, 1 (2009)

Methods of Creating Knowledge Graph by Linking Biological Databases

Nazar Zaki(✉), Chandana Tennakoon, Hany Al Ashwal, Alanoud Al Jaberi, and Amel Al Ameri

College of Information Technology, UAEU, Al Ain 15551, UAE
nzaki@uaeu.ac.ae

Abstract. A large number of biological databases are currently in use by scientists. These databases employ different formats, many of which can be converted into resource description format (RDF), which can be subsequently queried using semantic web methods. These databases have "inter" and "intra" database relationships. RDF has an inherent graph structure that facilitates exploration of connections between data via graphical representations known as knowledge graphs. In this paper, we survey the existing methods that are in use to link biological databases and evaluate the effectiveness with which the available approaches can predict unknown links between entities in databases as a means of improving knowledge graphs.

Keywords: Protein-protein interaction · Missing links
Information retrieval · Knowledge graph

1 Introduction

The scientific community currently makes use of thousands of different biological databases and they play a fundamental role in biological research. For basic entities in biology (e.g., genes, RNA, proteins, pathways, genetic variations, diseases, etc.) the databases that are available are typically well curated, cross-referenced, and documented. Although these databases are presented in diverse formats, we can transform them into the uniform RDF format [1]. Many notable databases can currently be found in RDF formats and a range of options are available by which it is possible to transfer databases of a variety of alternative formats into RDF. The RDF format has a natural graph structure associated with it, and each RDF statement describes a relationship between a subject and an object; that is, the predicate. The triplets of a subject, predicate, object form a graph with the subject and object as nodes and the predicate as a labeled, directed edge. An advantage of this graph structure is that any two nodes can be connected with an edge by simply adding an additional RDF statement. Some recognized databases are currently in RDF format and there are methods available to transform many other databases to RDF. The RDF format has a natural graph structure associated with it. Each RDF statement describes a relationship

© Springer Nature Switzerland AG 2019
F. Fdez-Riverola et al. (Eds.): PACBB 2018, AISC 803, pp. 52–62, 2019.
https://doi.org/10.1007/978-3-319-98702-6_7

between a subject and an object. The relationship itself is called the predicate. The triplets subject, predicate, object form a graph with subject and object as nodes and the predicate as a labeled, directed edge. An advantage of this graph structure is that any two nodes can be connected with an edge simply by the addition of an RDF statement. While isolated databases can be useful in their own rights, it would be much more beneficial if databases could be more readily linked so that related entries can be efficiently found and retrieved. For example, a user may be interested in identifying the diseases that are associated with a list of variations. If the variations database is linked with a database of diseases, he or she will be able to easily navigate through the intended results. Constructing such an interconnected knowledge graph would be relatively easy, provided we have the necessary details about how different entities in the databases are linked. Connecting two different databases can be trivial when related objects can be easily identified; for example, via a common identifier. However, when no such information is available, the user has to resort to inferring links. In addition to being helpful when attempting to connect two different databases, methods of inferring links can also be useful when we attempt to link elements with a database and when a database contains incomplete information.

Members of the biological community have engaged in extensive efforts to convert databases into RDF format and to subsequently link them. In this paper, we will present an overview of the ongoing studies that aim to integrate biological data. Furthermore, we will evaluate the methods that are available to infer links between these databases.

2   Linked Data Projects

Several projects have integrated biological data. Some such projects have been limited to a subset of biological data, while others have attempted to integrate biological data without any restrictions. In the proceeding sections, we will review some of the more significant linked data projects.

2.1   Bio2RDF

Bio2RDF [2] provided a set of custom scripts to download and convert a large set of databases. Around 35 databases currently use this system (including major databases such as UniProt, Entrez gene, and Kegg) and the scripts are compatible with sources that are in a range of formats, such as tab-delimited or XML. In addition, the data can be retrieved from SQL databases. Bio2RDF supports a federated data model and, since it is an open source project, users can host their own mirrors of the available data. For each database, bio2RDF assigns a namespace consisting of the preferred short name of data providers based on a resource registry. Data from the databases are assigned a unique URI of the form http://bio2RDF/namespace:ID, where the namespace is taken from the registry mentioned above and the ID is a data identifier that is provided by the database. For example, if we have a protein with UniProt ID

O14746, it would be assigned the URI http://bio2RDF/uniprot:O14746. If the source does not provide any identifying information, the URI is constructed as http://bio2RDF/namespace:namespace_resource:ID, where ID is a unique identifier that is generated by Bio2RDF using existing identifiers where appropriate. With these normalized URI, if new data are added that refers to an existing URI, they will be automatically linked together. It may be the case that a different database provider has already assigned a URI that is different to the normalized URI. In such instances, the two entities are linked using the predicate owl:sameAs. The predicates and data types in a database are represented using the convention http://bio2RDF/namespace_vocabulary:ID, where the ID can refer to a relation or information type. Bio2RDF can be used to create "mashups", where different data sources are combined to create a single data source. Bio2RDF facilitates the extraction of relevant data for creating such mashups, and examples of this process can be found in Belleau et al. [2].

2.2   Linked Life Data (LLD)

LLD [3] is a project that warehouses RDF data from different data sources. LLD is designed to systematically name URI's, given priority to the retention of the original RDF structure as distributed by the original distributor. Otherwise, LLD names databases in the format lld:resource/db/ and predicates in the format lld:resource/db/predicate. When a resource has a stable identifier, it is assigned a URI with the format lld:resource/db/type/id. To create linked data, LLD has identified several patterns of linkage between different databases. For example, a URI namespaseX:a and namespaceY:a may reference the same object and could be connected. Alternatively, we may be able to infer a transitive relationship from the tuples that X is related to Y and Y is related to Z, thereby implying that X could be related to Z. However, it is clear that these rules are not general and should be applied only to databases that are compatible with such restraints. In addition to automatically joining statements that share a common URI, as with Bio2RDF, LLD makes additional inferences using the graph structure and content of the RDF databases. LLD also uses information extraction processes to recognize and semantically annotate the text content of a database.

2.3   Linking Open Drug Data (LODD)

Linking Open Drug Data (LODD) [4] aims to link information about pharmaceuticals and holistic medical treatments. The linked information ranges from clinical trials to gene expressions and includes content from traditional Chinese medicine and pharmaceutical companies. Many commonly known identifiers in the life science domain can be utilized to link the data. A large number of these are already covered by the Bio2RDF project and have URIs for explicit linking. In more complex cases, LODD uses semantic link discovery and generation tools. Two of the main tools that are in use are as follows:

- LinQuer: A tool for semantic link discovery over relational data based on string and semantic matching techniques and their combinations. The Lin-Quer framework consists of LinQL, a declarative language that allows the user to specify linkage requirements across a wide variety of applications. The framework rewrites LinQL queries into standard SQL queries that can be run over existing relational data sources.
- Silk: An algorithm that discovers links between data sources by accessing the data using SPARQL. It provides the declarative Silk Link Specification Language (Silk-LSL) for specifying the link types and conditions. Link conditions apply similarity metrics, like string, numeric, data, URI, and set comparison methods, to entity properties. Link specifications allow only confident links above a given similarity threshold.

2.4   The Knowledge Base of Biomedicine (KaBOB)

The Knowledge Base of Biomedicine (KaBOB) [5] links databases using Open Biomedical Ontologies (OBOs). The entity identity is maintained across data sources through the generation of a single biomedical entity for each set of data source-specific identifiers that map to an entity. These entities, along with the OBO concepts, function as the building blocks for common biomedical representations. With the integration of biological ontologies, KaBOB enables the querying of data using biological concepts such as genes and processes. KaBOB initially starts with an empty triple store to which ontologies are directly added. Next, it adds RDF-converted data from various databases onto the triple store. Biomedical concepts in the ontologies are then connected to the RDF data.

2.5   Neurocommons

The Neurocommons [6] project predominantly aims to create a database that supports neurological disease research. When naming database objects, Neurocommons attempts to use existing URIs. If these are inadequate and fail the requirements for URIs applied by Neurocommons, new URIs are created. These new URIs take the form of persistent URLs (PURLs) for stability. Neurocommons generates additional data using text mining. PubMed is scanned for articles with abstracts and those articles related to neurology according to their MeSH (National Library of Medicine, 2016) annotation are retained. These abstracts are processed using a program called "Temis" in combination with the "biological entity recognizer" plugin. The result is then converted to RDF after further processing and subsequently added to the database.

3   Methods of Inferring Relationships in Knowledge Graphs

In this section, we will discuss the existing methods that can be used to infer new relations from a knowledge graph. These methods are divided into three main categories. In the following discussion let G be a graph with set of vertices V and edges E. For a vertex $v \in V$, let $N(v)$ denote the set of neighbors of v.

3.1   Observable Features

The first category of methods use observable features in the graph structure, and these approaches can be very useful when predicting links for single relationships. Examples of such single relationships would be protein-protein interaction analyses and friendships in social media. These methods are grounded in the principle of homophily [7], where we observe that similar entities have a higher chance of being related and, therefore, it is possible to derive the similarity of entities using the neighborhood of nodes or paths between nodes. Various features have been developed to capture these ideas; we review some of the popular graph features below. The common neighbors count the number of neighbors that two nodes $v_1, v_2 \in V$ share. For example, if v_1, v_2 were two individuals and the relationship, a large number of shared nodes $|N(v_1) \cap N(V_2)|$ between them would indicate they have lot of common friends. As such, we can then infer that there is a high likelihood that $(v_1, v_2) \in E$; i.e., that v_1, v_2 themselves are friends. Some of the main topological features that describe the edges, E, are as follows:

- Katz [8]: Measures a weighted sum of all paths between two vertices. Longer paths are penalized by assigning a path of length n an exponentially damped weight of β^n, where $\beta < 1$. The application of the Katz measure to the previous example states that friendship between v_1 and v_2 becomes less reliable as their degree of separation increases.
- Adar [9]: Assigns more importance to that exhibit "rare" connections, i.e. for $v_i \in N(v_1) \cap N(v_2)$ we assign a weight $\frac{1}{log(|N(v_i)|)}$ for v_i and add these weights. Adar is useful in cases where sharing rare features carry more information and where the rarity of connections lead to higher chance of interaction. In our example, a person with less friends is more likely to introduce his friends to one another than someone who has a high number of friends. Preferential attachment [11] is another feature that is somewhat opposite to Adar in that nodes having a large number of neighbors gets a higher score. For two nodes v_1 and v_2 it is easily calculated as $|N(v_1)||N(v_2)|$. This is useful as a similarity feature when nodes tend to connect with nodes having same preferences (e.g. sportsmen tend to connect with sportsmen).
- Jaccard [10]: Calculates the ratio $|N(v_1) \cap N(v_2)|/|N(v_1) \cup N(V_2)|$, i.e. the number of common neighbors per a neighbor of the two nodes v_1 and v_2. If this fraction of common neighbors is high, a high similarity is indicated.

Additional observable features can be calculated based on paths between two nodes. One of the most direct measures is the graph distance, which involves calculating the shortest path between two nodes. Hitting time for two nodes is the expected number of steps taken in a random walk between two nodes. Commute time [12] is its symmetric version, where the walk is made from one node to the other and back to the starting node. In rooted page rank [13], we calculate the probability that the node will be reached through a random walk, while randomly resetting the walk to a neighbor of the start node to avoid straying away in complicated paths. A variant of the rooted page rank is the Edge rank [14], where the rooted page rank is calculated over a weighted and undirected adjacency matrix. The power law exponent is the ratio of the number of edges to vertices in a sub-graph, taken in log scale. Betweenness centrality is another valuable feature that basically represents a method of identifying centrality, or the most important nodes in a graph using shortest paths. Some path-based calculations can be formulated as matrix problems. For example, the Katz measure can be calculated as a polynomial function of adjacency matrices. If $A(v_1, v_2)$ is the adjacency matrix of the nodes v_1 and v_2, $A^n(v_1, v_2)$ would give the number of paths connecting v_1 and v_2. Therefore the Katz measure can be written as $\sum_{n=1}^{\infty} \alpha^n A^n(v_1, v_2)$ for weight α. The exact form of these matrices maybe difficult to calculate for large graphs and approximations that are easier to calculate can be derive using singular-value decomposition (SVD) methods [9]. These features can be classified as local, global, and quasi-local features depending on the extent of the use of the graph. Local features such as Common Neighbors, Adar and Preferential Attachment use only their immediate neighborhood. Katz is a global measure; however, it can be transformed into a quasi-local measure by limiting the number of paths used to calculate it. A limitation of the local methods is that they might fail to capture patterns found across more distant neighbors.

3.2   Latent Features

Unlike the features previously described, these features are not outwardly observable but can be used to explain observable data. Algorithms have been developed that attempt to automatically infer latent features from data. One of the most effective of these methods is tensor factorization. A set of features E, and relationships R can be represented using a 3-way tensor X_{ijk}. A slice of this tensor is an $|E|$ by $|E|$ matrix and there are $|R|$ such slices. Let $|E| = n$ and $|R| = m$. The kth slices represent a relationship r_k, and we create an adjacency matrix by marking the element (i, j) with a 0 if the ith and j^{th} entities in E are related by r_k. We can decompose this tensor to the form $X_{ijk} = AR_k A^T$, where A is an $n \times r$ matrix that containing the latent variable representation of the entities. This factorization implies that the terms in X_{ijk} can be specified using r vectors. These are r latent variables and would give R_k as an $r \times r$ matrix that specifies the interaction of the latent components per predicate. Note that the final set of latent variables will not necessarily have a human readable meaning. To verify if there is a relation r_k between E_i and E_j, we calculate $a_i^T R_k a_j$, where a_i and a_j are the ith and j^{th} column vectors of A. If this quantity is larger than some θ the relationship is assumed to be correctly inferred. Several methods are available to factorize tensors. The standard methods are CP [15], PARAFAC [16] and TUCKER [17]. Many variants of these methods are also in use; for example CP-ALS,PARAFAC2 [18], sPARAFAC [19], PARALIND [20], and TUCKER2, TUCKER3, S-T3/S-T2 [21]. A more recent method, RESCAL [23] has proven to achieve a strong performance in comparizon to these alternative methods [22]. Following along the same idea, it is possible to use matrix factorization methods in place of tensor factorization. The $\mathbb{R}^n \times \mathbb{R}^n \times \mathbb{R}^m$ tensor is reshaped into a matrix with dimensions $\mathbb{R}^{n^2} \times \mathbb{R}^m$. The rows of this matrix consists of subject-object pairs (p, q) while the columns are relations r_k. Alternatively, the tensor can be converted to a $\mathbb{R}^n \times \mathbb{R}^{nm}$ matrix that consists of rows with subjects and columns with relation-object pairs. Compared to tensor factorization, these methods cannot completely integrate the subject-relation-object data. For example, if each (subject, object) pair is modeled together, the information about the relationships that share common objects will not be captured. In addition, studies have found that matrix representations consume more memory [24]. Jiang et al. [25] and Riedel et al. [26] described how matrices can be reshaped using subject-object pairs, while Huang et al. [27] and Tresp et al. [28] described how they can be performed using relation-object pairs. These methods have the advantage of using several relationships when inferring new relationships. They can also infer many relationships at the same time in comparison to observable variable methods.

3.3   Markov Random Fields

Markov random fields work on the assumption that nodes in a graph have local interactions. A Markov random field is a collection of random variables represented by an undirected graph that has a Markov property. Given a knowledge graph, we can construct a corresponding "dependency graph" in which random variables are the relationships, and the dependencies between two such random variables are represented by the edge that connects them. The structure of the dependency graph is defined using the Markov logic language. For example, if the relationships x and y are parents of z, using Markov logic, we can specify that x and y are married and z is not married to either x or y using Markov logic [29]. There are several methods available to infer probable relationships using this dependency graph, Gibbs sampling [30] being the most prominent. Table 1 shows the pros and cons in the methods discussed in this section.

Table 1. Comparison of characteristics of different methods used to infer links.

	Uses local information only	Fast	Suitable for large databases	Easy to automate	Single relations Only
Local methods	Yes	Yes	Yes	Yes	Yes
Global methods	No	No	No	Yes	Yes
Quazi-local methods	No	Yes	Sometimes	Yes	Yes
Markov random fields	No	No	No	Yes	No
Latent variables	No	Yes	Yes	No	No

4 Comparison Between Different Approaches

In this section, we compare the methods described above and evaluate which are the most suitable for predicting links in biological databases. Since the database sizes are generally very large, we need methods that can scale well. Furthermore, we are seeking a link prediction procedure that can be automated as much as possible. Calculating features using observable features is generally fast for local similarity indices. The use of global similarity measures will not be practical for very large databases; as such quasi-local methods in which calculations are performed with bounded paths are a better option. Efficient tensor factorization methods such as RESCAL [31] have reported fast link prediction times on large data sets. The exact solution of the Markov random fields is very time consuming, and would not be suitable on the genomic database scale. Graph feature calculation and generating the 3-way tensor for latent variable analysis will not pose a problem in automation. The calculations and the process by which the tensor structure is created are straightforward. However, with Markov models the dependency graphs would need to be specified on a case by case basis. As such, it is not particularly suitable for automation. The observable feature methods tend to perform well when a single relationship is being inferred. With the latent feature methods, no such restriction is noticed.

We conducted a test under several machine learning algorithms implemented in Weka to determine the extent to which different methods could effectively predict links under several machine learning algorithms. We generated a random graph as shown in Fig. 1, and calculated the local, quasi-global and global features for each pair of nodes. The graph contains 23 nodes and 29 edges. The graph density is 0.115, the clustering coefficient is 0.623, the shortest path is 122 (22%) and the average number of neighbors is 2.522. Based on the values of these features, we ran the classification algorithms Support vector machines (SVM), k-nearest neighbor (IBK), decision tree (DT) and Bayes Net (BN) to classify pairs of nodes into two classes; linked nodes and nodes without a link.

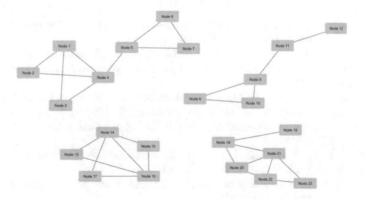

Fig. 1. A simulated graph used for predicting links

Table 2 shows the results of the link prediction on the simulated graph. With the exception of the IBK method, the precision (PR), accuracy (AC) and recall (RE) are increased as more global information is utilized. In fact, the top two algorithms (SVM and IBK) achieved 100% PR, RE and AC when global information were used in features.

Table 2. Link prediction performance of different methods under several classification algorithms.

Classifiers	Local features			QuaziGlobal features			Global features		
	PR	RE	AC	PR	RE	AC	PR	RE	AC
SVM	0.751	0.75	0.75	0.887	0.875	0.875	1	1	1
IBK	0.902	0.893	0.8929	0.821	0.821	0.8214	1	1	1
DT	0.838	0.821	0.8214	0.844	0.839	0.8393	0.983	0.982	0.9821
BN	0.556	0.554	0.5538	0.873	0.857	0.8571	0.982	0.982	0.9821

5   Conclusion

In this paper, we reviewed the existing methods that are available to link biological databases with a focus on methods to predict unknown links between entities (nodes) in the databases. We discussed and evaluated three categories of features used in conjunction with machine learning classifiers to detect possible missing links that could improve the knowledge graph. Future research will employ several real graph data sets to evaluate the mentioned features in addition to employing feature selection techniques to improve detection accuracy.

References

1. Beckett, D., McBride, B.: RDF/XML Syntax Specification. W3C Work, pp. 1–56 (2003)
2. Belleau, F., et al.: Bio2RDF: towards a mashup to build bioinformatics knowledge systems. J. Biomed. Inform. **41**(5), 706–716 (2008)
3. Momtchev, V., Peychev D., Primov, T., Georgiev, G.: Expanding the pathway and interaction knowledge in linked life data. In: Proceedings of International Semantic Web Challenge (2009)
4. Samwald, M., et al.: Linked Open drug data for pharmaceutical research and development. J. Cheminfo. **3** (2011). https://doi.org/10.1186/1758-2946-3-19
5. KaBOB: ontology-based semantic integration of biomedical databases. BMC Bioinform. **16**, 126 (2015)
6. Ruttenberg, A., et al.: Life sciences on the semantic web: the neurocommons and beyond. Brief. Bioinfo. **10**, 193–204 (2009)
7. Lauw, H., et al.: Homophily in the digital world: a live journal case study. IEEE Internet Comput. **14**, 15–23 (2010)
8. Katz, L., et al.: A new status index derived from sociometric analysis. Psychometrika **18**, 39–43 (1953)

9. Acar, E., et al.: Link prediction on evolving data using matrix and tensor factorizations. In: 2009 IEEE International Conference on Data Mining Workshops, pp. 262–269 (2009)
10. Jaccard, P., et al.: Étude comparative de la distribution florale dans une portion des Alpes et des Jura. Bull. del la Société Vaudoise des Sci. Nat. **37**, 547–579 (1901)
11. Newman, M.E.J., et al.: Clustering and preferential attachment in growing networks. Phys. Rev. **64**, 25102 (2001)
12. Liu, W., Lu, L.: Link prediction based on local random walk. EPL (Europhysics Lett.) **89**, 58007 (2010)
13. Liben-Nowell, D., Kleinberg, J.: The link prediction problem for social networks. In: Proceedings of Twelfth Annual ACM International Conference Information and Knowledge Management, pp. 556–559 (2003)
14. Brin, S., Page, L.: The anatomy of a large-scale hypertextual Web search engine BT, Computer Networks and ISDN Systems. Comput. Netw. ISDN Syst. **30**, 107–117 (1998)
15. Carroll, J.D., Chang, J.J.: Analysis of individual differences in multidimensional scaling via an n-way generalization of 'Eckart-Young' decomposition. Psychometrika **35**, 283–319 (1970)
16. Harshman, R.: Foundations of the PARAFAC procedure: models and conditions for an 'explanatory' multimodal factor analysis. UCLA Work. Pap. Phonetics **16**, 1–84 (1970)
17. Tucker, L.R.: The extension of factor analysis to three-dimensional matrices. In: Contributions to Mathematical Psychology, pp. 110–119 (1964)
18. Tucker, L.R.: PARAFAC2: mathematical and technical notes. UCLA Work. Pap. Phonetics **22**, 30–44 (1972)
19. Hong, S.J., Harshman, R.: Shifted factor analysis, Part III: N-way generalization and application. J. Chemom. **17**, 389–399 (2003)
20. Bro, R., et al.: Modeling multi-way data with linearly dependent loadings. J. Chemom. **23**, 324–340 (2009)
21. Harshman, R.A., et al.: Shifted factor analysis? Part I: models and properties. J. Chemom. **17**, 363–378 (2003)
22. Nickel, M., et al.: Factorizing YAGO. In: Proceedings of the 21st international conference on World Wide Web - WWW 2012, p. 271 (2012)
23. Nickel, M., Tresp, V.: Tensor factorization for multi-relational learning. Lecture Notes in Computer Science, pp. 617–621 (2013)
24. Nickel, M., et al.: A review of relational machine learning for knowledge graphs. Proc. IEEE **104**, 11–33 (2016)
25. Jiang, X., et al.: Link prediction in multi-relational graphs using additive models. In: International Workshop on Semantic Technologies meet Recommender Systems and Big Data at the ISWC, pp. 1–12 (2012)
26. Riedel, S., et al.: Relation extraction with matrix factorization and universal schemas. In: Proceedings 2013 Conference of the North American Chapter of the Association Computational Linguistics Human Language Technologies, pp. 74–84 (2013)
27. Huang, Y., et al.: A scalable approach for statistical learning in semantic graphs. Semantic Web **5**, 5–22 (2014)
28. Tresp, V., et al.: Materializing and querying learned knowledge. In: CEUR Workshop Proceedings (2009)
29. Richardson, M., Domingos, P.: Markov logic networks. In: Machine Learning, pp. 107–136 (2009)

30. Bishop, C.M.: Pattern Recognition and Machine Learning. Springer (2006). ISBN 0-387-31073-8
31. Nickel, M., Tresp, V., Kriegel, H.P.: A three-way model for collective learning on multi-relational data. In: Proceedings of the 28th International Conference on Machine Learning, pp. 809–816 (2011)

QSAR Modelling for Drug Discovery: Predicting the Activity of LRRK2 Inhibitors for Parkinson's Disease Using Cheminformatics Approaches

Víctor Sebastián-Pérez[1], María J. Martínez[2], Carmen Gil[1],
Nuria E. Campillo[1], Ana Martínez[1], and Ignacio Ponzoni[2(✉)]

[1] Centro de Investigaciones Biológicas (CIB, CSIC),
Ramiro de Maeztu 9, 28040 Madrid, Spain
[2] Instituto de Ciencias e Ingeniería de la Computación (UNS–CONICET),
Departamento de Ciencias e Ingeniería de la Computación,
Universidad Nacional del Sur (UNS), Bahía Blanca, Argentina
ip@cs.uns.edu.ar

Abstract. Parkinson's disease is one of the most common neurodegenerative disorders in elder people and the leucine-rich repeat kinase 2 (LRRK2) is a promising target for its pharmacological treatment. In this paper, QSAR models for identification of potential inhibitors of LRRK2 protein are designed by using an in house chemical library and several machine learning methods. The applied methodology works in two steps: first, several alternative subsets of molecular descriptors relevant for characterizing LRRK2 inhibitors are identified by a feature selection software tool; secondly, QSAR models are inferred by using these subsets and three different methods for supervised learning. The performance of all these QSAR models are assessed by traditional metrics and the best models are analyzed in statistical and physicochemical terms.

Keywords: Cheminformatics · QSAR · Machine learning
Parkinson's disease · LRRK2

1 Introduction

Nowadays, the search of effective treatments for neurodegenerative diseases is one of the urgent clinical and social needs. Number of people affected by those pathologies, including Alzheimer's and Parkinson's diseases, increase every year, mainly in developed countries, directly associated to the longer life expectancy. Parkinson's Disease (PD) is the second most common human neurodegenerative disorder in people over 60 years of age, affecting 1 in 100 people and increasing to that affects 2–3% of the population ≥ 65 years of age. It is associated with Lewy bodies, abnormal aggregates of α-synuclein protein, and loss of dopaminergic neurons in the substantia nigra. Although clinical diagnosis is based on the existence of bradykinesia and other cardinal motor characteristics, Parkinson disease is associated with many non-motor symptoms that add to overall disability.

© Springer Nature Switzerland AG 2019
F. Fdez-Riverola et al. (Eds.): PACBB 2018, AISC 803, pp. 63–70, 2019.
https://doi.org/10.1007/978-3-319-98702-6_8

Epidemiological and genetic studies carried on several families in Asia, the United States, and Europe led to discover in 2004 that mutations in a new gene, known as leucine-rich repeat kinase 2 (LRRK2), are a major genetic risk factor for familiar and sporadic PD [1]. Today, LRRK2 is one of the most pursuing and promising targets for the future pharmacological treatment of PD. In this sense, big efforts are being done both from academia and pharmaceutical industry with the goal of developing selective and brain-permeable LRRK2 inhibitors as a strategy for PD [2, 3]. LRRK2 is an unusual large protein (2527 amino acids) classified as a member of the ROCO superfamily. It presents a leucine-rich repeat (LRR) domain, a kinase domain, a DFG-like motif, a RAS domain, a GTPase domain, a MLK-like domain, and a WD40 domain. The protein is present mainly in the cytoplasm, although it is also related to the mitochondrial outer membrane. The physiological role of LRRK2 is poorly understood and many of its substrates remain unclear. However, it has been proposed to be beneficial for preventing neurodegeneration [4, 5] and several LRRK2 inhibitors are being developed as neuroprotective agents for PD. Some studies revealed that LRRK mutations increases aggregation of α-synuclein in dopaminergic neurons that are exposed to α-synuclein fibrils [6].

Quantitative structure–activity/property relationship (QSAR/QSPR) modeling has been established itself as one of the major computational molecular modeling methodologies, playing a central role in drug identification or optimization. QSAR/QSPR models allow to identify relationships between structural information of chemical compounds (molecular descriptors) and a physicochemical or biological property under study. Now, these techniques are widely used as a surrogate for experimental studies to predict the activity of the molecules from their structure. In particular, machine learning methods had become extensively used in this field during the last decades [7].

Regarding literature, very few QSAR studies in LRRK2 have been published. The works that have been reported presented a very limited predictive activity for the external validation datasets [8, 9]. In this paper, novel QSAR models for predicting putative inhibitors of LRRK2 protein are developed by using machine learning methods. In particular, several regression and classification QSAR models are inferred and their performances contrasted in terms of accuracy and model complexity.

2   Materials and Methods

Several QSAR models inferred by feature selection techniques, both for classification and regression models, are described. Figure 1 presents a scheme of the experimental design followed in this work. The database is a compilation of 67 compounds previously synthesized in our research group and tested as inhibitors of LRRK2 enzyme [10]. In this assay, LRRK2 kinase activity was measured as the percentage of enzyme inhibition for every compound as it is further discussed in the previous reference. This information is available on http://lidecc.cs.uns.edu.ar/pacbb2018.LRRK2/structures_activity.html.

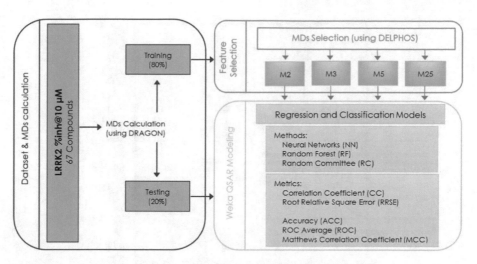

Fig. 1. Scheme of the *in silico* experiments reported for predicting the activity of LRRK2 inhibitors.

Database Analysis and Drug-Like Properties Calculation

A crucial step in QSAR studies is to collect a representative set of compounds in order to include a diversity physicochemical space. With the aim of analyzing the dataset, we have performed a characterization of the compounds, both from a physicochemical perspective and a drug-like point of view. Physicochemical and drug-like properties of this dataset were calculated using Qikprop and the most representative descriptors were analyzed to show the diversity of the dataset.

Some of these parameters are plotted in Fig. 2, where it can be observed a wide dispersion of 2 different key properties in drug discovery such as logP (x axis) and H-bond acceptors (accptHB y-axis). Compounds are colored taking into account their stars values. This parameter represents the number of properties or descriptors values calculated that fall outside the 95% range of similar values for known drugs. For this reason, a large value of stars suggests that a molecule is less drug-like. In this case, all compounds present a value equal or lower than 2, which means that the complete database is based on drug-like structure. Furthermore, taking into account Lipinski rule of 5 [11], the 67 compounds present 1 or none violations of the rules, which means that the molecules have properties that make them likely orally active drug in humans. Therefore, and after the analysis carried out, we can conclude that the database is diverse in terms of physicochemical properties and all the compounds are drug like.

QSAR Models

A total number of 3224 descriptors were computed using Dragon for the entire database. The experiments were designed following the procedure described in Fig. 1. The dataset was first divided into training (75%) and test set following a stratified sampling. Several subsets of descriptors were selected from the training set using DELPHOS [13]. This tool repeats ten times the random partition of the data in 75/25 to perform the validation of the selected characteristics. In Table 1, a summary of the best molecular

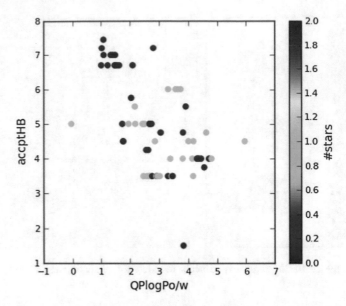

Fig. 2. Dispersion of the database regarding PQ properties, colors are defined by stars.

Table 1. Best molecular descriptors subsets obtained by DELPHOS feature selection tool.

Subset	Cardinality	MDs		Descriptor type
M2	4	MW		Constitutional indices
		MWC08		Walk and path counts
		BEHp2		Burden eigenvalues
		RDF105p		RDF descriptors
M3	4	MW		Constitutional indices
		JGI2		2D autocorrelation
		HATs6m	R2e	GETAWAY descriptors
M5	5	MW		Constitutional indices
		IC0		Information indices
		ESpm09x		Edge adjacency indices
		JGI3		2D autocorrelation
		L3s		WHIM descriptors
M25	13	MW		Constitutional indices
		HNar	ECC	Topological indices
		GATs7e		2D autocorrelations
		VEZ1	VEp2	2D matrix-based descriptors
		DISPm		Geometrical descriptors
		RDF105p		RDF descriptors
		R8e		GETAWAY descriptors
		B06[N-Br]	B07[C-Cl]	2D atom pairs
		F04[C-C]	F05[O-Cl]	2D atom pairs

descriptors subsets in terms of RAE (Relative Absolute Error) is reported. Using these 4 subsets and different inference methods, a variety of QSAR models were built for regression and classification. The models are computed by WEKA [14] using Neural Networks (NN), Random Forest (RF) and Random Committee (RC) as inference methods, for computing these models default parameters were used in WEKA. Several methods for obtaining the QSAR models were tested due to the fact that recent studies have shown that there does not exist a more advisable strategy for inferring the QSAR from the subsets of descriptors [12]. For classification models, discretization thresholds of target property values were as follows: low activity $\leq 50\%$ and high activity $>50\%$. Table 2 shows several metrics computed using WEKA for the best regression and classification QSAR models obtained with each descriptor subset. The performance results for classification models are reported using the accuracy (ACC), namely percentage of cases correctly classified, the average Receiver Operating Characteristic (ROC) and the Matthews Correlation Coefficient (MCC). For regression models, the correlation coefficient (CC) and the root relative square error (RRSE) results are informed.

Table 2. Performances of the best regression and classification QSAR models for the external validation testing set.

Model	Cardinality	Best regression QSAR models			Best classification QSAR models			
		Method	CC	RRSE	Method	ACC	ROC	MCC
M2	4	RF	0.55	87.50	RF	68.8	0.69	0.40
M3	4	RC	0.68	74.69	RC	87.5	0.90	0.77
M5	5	RF	0.83	60.82	RC	75.0	0.95	0.52
M25	13	RC	0.44	92.34	RC	75.0	0.73	0.53

The best classification model was inferred from the subset M3 by using Random Committee and achieved 87.5% of correct classification with a ROC value of 0.91, all results are shown in Table 2. In other hand, the best regression model was obtained with the subset M5 by using Random Forest and achieved a correlation coefficient of 0.83 and a RRSE of 60.82. The datasets used to generate the classification and regression models can be found in link http://lidecc.cs.uns.edu.ar/pacbb2018.LRRK2/datasets.html.

The best model found in the regression case contains 5 different descriptors that includes a wide variety of different descriptors classes, from 0D molecular descriptors to 3D. For example, molecular weight is a constitutional index, inside the class of 0D-descriptors, that is obtained from the chemical formula, as they do not consider the tridimensional structure of the ligands. We have also found ESpm09x descriptor, spectral moment 09 from edge adjacency matrix weighted by edge degrees and IC0 information content index (neighborhood symmetry of 0-order) and JGI3: mean topological charge index of order 3 that are 2D descriptors from different families. Finally, the 3D descriptor found in this model is the L3s: 3rd component size directional WHIM index/weighted by I-state. Regarding the best classification model, includes 4 descriptors and we have also found a wide representation of different descriptors families. MW has also been chosen in this model as a 0D descriptor. A very similar descriptor to JGI3 was also found in this model, JGI2, is a 2D descriptor and is related to the charge index of the compounds in the database. Finally, two 3D descriptors were selected by DELPHOS that are R2e: R autocorrelation of lag 2/weighted by Sanderson electronegativity and HATs6m: leverage-weighted autocorrelation of lag 6/weighted by mass. Both are GETAWAY descriptors (GEometry, Topology, and Atom-Weights AssemblY) and are related to electronegativity and mass, key parameters in protein-ligand recognition process for protein inhibition.

Furthermore, we analyzed the relationship among the descriptors in statistical terms by using VIDEAN, which is a visual analytics tool for the study of molecular descriptor subsets. It shows the relationships and interactions between the descriptors and the target property in statistical terms. The analysis of the pair correlation between the five descriptors that conform the regression model and the four descriptors that are part of the best classification model is presented in Fig. 3 using Kendall tau metric.

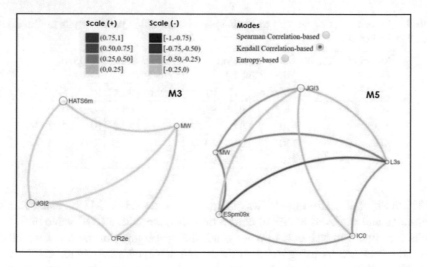

Fig. 3. Kendall correlation analysis among the descriptors that conforms the subsets M3 y M5.

In this type of analysis, the main goal is to identify models with low correlation among descriptors that means low redundant information. In this case, we observe a clear majority of light tones of links, both blue and orange, that make connection between the descriptors (nodes) present in the models. This fact, demonstrate the low data redundancy in both models that have been selected as the best ones.

All the experimental section and protocols regarding database analysis and property calculation as well as QSAR models both building and analysis can be found at: http://lidecc.cs.uns.edu.ar/pacbb2018.LRRK2/supplementary_material.html.

3 Conclusions

Parkinson's disease constitutes one of the neurodegenerative disorders with higher impact in elder population around the world. An auspicious target for its pharmacological treatment is the leucine-rich repeat kinase 2 (LRRK2). Several studies proposed that LRRK2 inhibitors can be a beneficial strategy for preventing neurodegeneration. For this reason, in this paper, QSAR models have been developed with the aim to use them as useful filters for virtual screening to identify potential inhibitors of LRRK2 protein. These models were obtained by machine learning methods over data from an in house chemical library.

The computational approach used in this work follows two main steps: first, alternative subsets of molecular descriptors relevant for structural characterization of LRRK2 inhibitors are identified by a feature selection method; secondly, QSAR models are learned from these subsets by applying several supervised learning algorithms. The performance of these QSAR models was contrasted by traditional metrics.

The molecular descriptor subsets associated with the regression and classification models that reached the best performances were analyzed in statistical and physicochemical terms. From the analysis, it is possible to observe that the selected subset has low cardinality but cover a wide spectrum of the molecular descriptor classes, contributing in this way with meaningful and diverse structural information to the models. Besides, the visual analytics study reveals that the selected molecular descriptors provides non-redundant information in statistical terms.

Nevertheless, even when these QSAR models achieve high accuracies, it is important to mention that these models have been learned from datasets integrated by a reduced number of chemical compounds, which can limit the generalization properties of these predictive models. For this reason, our advice for potential practitioners of these models is to employ applicability domain methods over their testing compounds before apply these models. As future work, we hope to extend our in house chemical library for LRKK2 in order to improve the generalizability of these achievements.

Acknowledgments. This work is kindly supported by CONICET, grant PIP 112-2012-0100471 and UNS, grants PGI 24/N042 and PGI 24/ZM17. We also acknowledge MECD, VSP grant FPU15/01465 and Banco Santander for VSP fellowship AY21/17-D-27 in the "Becas Iberoamerica-Santander Investigación" program.

References

1. Zimprich, A., Biskup, S., Leitner, P., Lichtner, P., Farrer, M., Lincoln, S., Kachergus, J., Hulihan, M., Uitti, R.J., Calne, D.B., Stoessl, A.J., Pfeiffer, R.F., Patenge, N., Carbajal, I.C., Vieregge, P., Asmus, F., Muller-Myhsok, B., Dickson, D.W., Meitinger, T., Strom, T.M., Wszolek, Z.K., Gasser, T.: Mutations in LRRK2 cause autosomal-dominant parkinsonism with pleomorphic pathology. Neuron **44**, 601–607 (2004)
2. Gilligan, P.J.: Inhibitors of leucine-rich repeat kinase 2 (LRRK2): progress and promise for the treatment of Parkinson's disease. Curr. Top. Med. Chem. **15**, 927–938 (2015)

3. Estrada, A.A., Sweeney, Z.K.: Chemical biology of Leucine-Rich repeat Kinase 2 (LRRK2) inhibitors. J. Med. Chem. **58**, 6733–6746 (2015)
4. Cookson, M.R.: LRRK2 pathways leading to neurodegeneration. Curr. Neurol. Neurosci. Rep. **15**, 42 (2015)
5. Smith, W.W., Pei, Z., Jiang, H., Moore, D.J., Liang, Y., West, A.B., Dawson, V.L., Dawson, T.M., Ross, C.A.: Leucine-rich repeat kinase 2 (LRRK2) interacts with parkin, and mutant LRRK2 induces neuronal degeneration. Proc. Natl. Acad. Sci. U.S.A. **102**, 18676–18681 (2005)
6. Volpicelli-Daley, L.A., Abdelmotilib, H., Liu, Z., Stoyka, L., Daher, J.P., Milnerwood, A.J., Unni, V.K., Hirst, W.D., Yue, Z., Zhao, H.T., Fraser, K., Kennedy, R.E., West, A.B.: G2019S-LRRK2 expression augments alpha-synuclein sequestration into inclusions in neurons. J. Neurosci. **36**, 7415–7427 (2016)
7. Lima, A., Philot, E., Trossini, G., Scott, L., Maltarollo, V., Honorio, K.: Use of machine learning approaches for novel drug discovery. Expert Opin. Drug Discov. **11**, 225–239 (2016)
8. Kahn, I., Lomaka, A., Karelson, M.: Topological fingerprints as an aid in finding structural patterns for LRRK2 inhibition. Mol. Inform. **33**, 269–275 (2014)
9. Pourbasheer, E., Aalizadeh, R.: 3D-QSAR and molecular docking study of LRRK2 kinase inhibitors by CoMFA and CoMSIA methods. SAR QSAR Environ. Res. **27**, 385–407 (2016)
10. Salado, I.G., Zaldivar-Diez, J., Sebastian-Perez, V., Li, L., Geiger, L., Gonzalez, S., Campillo, N.E., Gil, C., Morales, A.V., Perez, D.I., Martinez, A.: Leucine rich repeat kinase 2 (LRRK2) inhibitors based on indolinone scaffold: Potential pro-neurogenic agents. Eur. J. Med. Chem. **138**, 328–342 (2017)
11. Lipinski, C.A., Lombardo, F., Dominy, B.W., Feeney, P.J.: Experimental and computational approaches to estimate solubility and permeability in drug discovery and development settings. Adv. Drug Deliv. Rev. **46**, 3–26 (2001)
12. Eklund, M., Norinder, U., Boyer, S., Carlsson, L.: Choosing feature selection and learning algorithms in QSAR. J. Chem. Inf. Model. **54**, 837–843 (2014)
13. Soto, A.J., Martínez, M.J., Cecchini, R.L., Vazquez, G.E., Ponzoni, I.: DELPHOS: computational tool for selection of relevant descriptor subsets in ADMET prediction. In: 1st International Meeting of Pharmaceutical Sciences (2010)
14. Frank, E., Hall, M.A., Witten, I.H.: The WEKA Workbench. Online Appendix for "Data Mining: Practical Machine Learning Tools and Techniques", 4th edn. Morgan Kaufmann, Massachusetts (2016)

Improving the Use of Deep Convolutional Neural Networks for the Prediction of Molecular Properties

Niclas Ståhl[1]([⊠]), Göran Falkman[1], Alexander Karlsson[1], Gunnar Mathiason[1], and Jonas Boström[2]

[1] School of Informatics, University of Skövde,
Högskolevägen 28, 54145 Skövde, Sweden
`niclas.stahl@his.se`
[2] Department of Medicinal Chemistry, CVMD iMED, AstraZeneca,
Pepparedsleden 1, 43183 Mölndal, Sweden

Abstract. We present a flexible deep convolutional neural network method for the analyse of arbitrary sized graph structures representing molecules. The method makes use of RDKit, an open-source cheminformatics software, allowing the incorporation of any global molecular (such as molecular charge) and local (such as atom type) information. We evaluate the method on the Side Effect Resource (SIDER) v4.1 dataset and show that it significantly outperforms another recently proposed method based on deep convolutional neural networks. We also reflect on how different types of information and input data affect the predictive power of our model. This reflection highlights several open problems that should be solved to further improve the use of deep learning within cheminformatics.

Keywords: Graph convolutional neural network
Molecular property prediction · Drug discovery

1 Introduction

Discovering new chemical entities is a difficult and time-consuming endeavour. By using computers and algorithms the belief is to shorten the process from idea to launched drug. The field of cheminformatics is one area holding promise to enable faster and better decision-making, for example, by deriving new methods for automated predictions of molecular properties [16]. Due to the nature of these problems, cheminformatics is influenced by the development and trends in machine learning. One sub-field in machine learning that has expanded tremendously in the last years, mainly due to the success in several fields such as image recognition and speech recognition, is *deep learning* [10]. Recently deep learning has started to make its way into cheminformatics [3].

© Springer Nature Switzerland AG 2019
F. Fdez-Riverola et al. (Eds.): PACBB 2018, AISC 803, pp. 71–79, 2019.
https://doi.org/10.1007/978-3-319-98702-6_9

Several traditional machine learning algorithms have been applied for the prediction of molecular properties, such as *support vector machines* and *artificial neural networks* [3,16]. A major difficulty when applying machine learning algorithms on molecule data is that most algorithms require input of a fixed size while molecules are of arbitrary size. To circumvent this problem, most authors pre-process their molecular data into chemical "fingerprints" enabling the application of selected machine learning algorithms. This process is for example used in [5,13]. However, some information about the molecular structure is lost in this step. In this paper we take a different approach and use *deep convolutional neural networks (DCNN)* on molecules represented as graphs. The same approach has been used by several authors [2,6,19]. We show that the predictive power of such networks can be improved if more information about the atoms, bonds and molecules are added, for example chirality, bond type and the molecular mass. While some, but not all, of the information above have been used in previously presented DCNN models, a reflection on how the selection of input information affects their result is missing. To show that this cannot be overlooked we develop a flexible model where both global and local information can be easily incorporated or removed. Using this model we explore how the predictive power varies when more information is gradually incorporated. It is worth to point out that this study does not aim to be an extensive search to find which information that adds the most to the predictive power. Instead, we aim to show that different types of information about molecules easily can be incorporated into a DCNN and that this can increase the predictive power of that particular network.

There has not been any standardised way to measure cheminformatics algorithms against each other. However most recently Wu et al. [19] made an effort and compiled several datasets that can be used as benchmarks for this purpose. In this paper we select one specific dataset from [19] to work with; the SIDER dataset which originates from the SIDER database [8]. The reason for why we selected this dataset is that Wu et al. [19] acknowledge this to be a dataset where it is difficult to achieve a high predictive accuracy. Surprisingly, while being one of the best performing methods on all other datasets in [19] their presented DCNN performed very poorly on the SIDER dataset and were outperformed by traditional methods such as *random forest* and *logistic regression*. However, both of these methods are less flexible and require the use of chemical fingerprints. Therefore we argue that there is a great need for the improvement of DCNN methods to be used on this dataset. Another reason for why this dataset is particularly interesting is that no adequate investigation was done in order to find out why the performance of the DCNN was particularly poor on this dataset. By further understanding the reasons behind this, we hope to show how deep learning methods can be improved in the future.

2   Deep Learning in Cheminformatics

Neural networks have been used in cheminformatics for several decades and several authors have applied deep neural networks on molecular fingerprints for the prediction of many different properties, including toxicity [13] and solubility [5]. One of the most famous applications of neural networks in cheminformatics is the work of Dahl et al. [1], which won the Merck Molecular Activity Challenge. However, such deep neural networks can only be applied to molecular fingerprints and hence, feature extraction must be conducted to reduce molecules into a fixed set of values. Several authors overcome this flaw by either using a recurrent neural network (RNN) or a DCNN. Lusci et al. [11] do for example use a RNN for the prediction of aqueous solubility. Another approach, that has been used by other authors, is to use a DCNN for the prediction of molecular properties. These authors assume that low level features in the molecule will emerge due to local interactions between neighbouring atoms in the same way as low level image features, such as edges, emerge due to interactions between neighbouring pixels in an image. Wallach et al. [17] for example apply a three dimensional DCNN on the spatial structure of molecules. Duvenaud et al. [2], Kearnes et al. [6] and Wu et al. [19] use another approach and represent molecules as graphs where nodes correspond to atoms and the edges to bonds. These DCNNs, which use different types of molecular information and with different purposes, are then applied to graphs representing molecules. Duvenaud et al. [2] for example use a DCNN that uses atom properties, such as the hybridization type of the atom and if the atom is in a ring or not, and bond properties such as the type of the bond to automatically generate molecular fingerprints. Kearnes et al. [6] use a different type of DCNN, which consists of "weave modules", where information is transferred between different atoms. Wu et al. [19] present several chemical datasets for benchmarking together with the predictive performance of several algorithms, among them a DCNN for graphs. However, even though these authors do not use the same atom and bond information in their networks, a reflection on how this affects the results is missing.

3   Proposed Model

The presented model architecture is a slightly modified version of the DCNN presented by Duvenaud et al. [2]. The first step is to calculate the initial representation of each atom. Let $A_i^{(0)}$ be a vector consisting of all information that is extracted from atom i. Here the superscript represents the layer number. The contents of this vector varies between the different experiments, described in Sect. 4.2. In the first and most simplistic experiment, $A_i^{(0)}$ corresponds to an "one hot" encoded vector, that is a vector where each elemental type corresponds to a given position in the vector. The value at the position of the elemental type of atom i is 1 and the value of all other positions are 0. In later experiments, $A_i^{(0)}$, the "one hot" encoded vector, is appended with real values, representing attributes such as the number of other atoms atom i binds to. Let $A_i^{(1)}$ be the vector representing the first hidden state of atom i, given by

$$A_i^{(1)} = \mathcal{F}\left(A_i^{(0)} \; ; \; W^{(0)}\right),$$

(1)

where $\mathcal{F}$ is an arbitrary activation function and $W^{(0)}$ is its parameters. Due to the conventions of the field, $\mathcal{F}$ will be defined as the leaky ReLU function [12] in the rest of this paper. The next operation in the model is the convolutional steps. In these steps, information is transmitted between atoms and their neighbours. The steps can be expressed as

$$C_{i,j}^{(l)} = \mathcal{G}^{(l)} \left(A_i^{(l)} \frown \mathcal{B}(i,j) \frown A_j^{(l)} \; ; \; W^{(l)} \right) \tag{2}$$

$$A_i^{(l+1)} = \left[\max_{j \in neighbourhood(i)} C_{i,j,1}^{(l)}, \ldots, \max_{j \in neighbourhood(i)} C_{i,j,k}^{(l)} \right] \tag{3}$$

Equation (2) calculates how atom i is affected by having atom j in its neighbourhood. The function $\mathcal{B}$ in Eq. (2) extracts information about the interactions between atom i and j, for example the type of bond between them. The $\frown$ operator represents concatenation of two vectors. $\mathcal{G}^{(l)}$ is an arbitrary activation function, and as with the function $\mathcal{F}$, it will be defined as the leaky ReLU function in the rest of this paper. The hidden representation in each atom is then updated as described by Eq. (3). Here the max function is applied element-wise to every element in the vectors and the results are then multiplied with the square matrix $V^{(l)}$ which is one of the parameters that will be learnt by the model, following the method used by He et al. [4]. This result is added to the previous hidden representation of atom i. This architecture is inspired by the work of He et al. [4], which showed that DCNNs having short-cut connections, a connection where the signal is passed forward without any modification, were more stable than those which did not have short-cut connections. Notice that there is nothing preventing $A_i^{(l)}$ and $A_i^{(l+1)}$ to be vectors of different sizes. However, for simplicity, we have decided to keep the size of the hidden representation fixed in all layers in our model. After m convolutional and max-pooling steps we find a global representation for the molecule. This is done by summing up the representation of all atoms and also adding global molecular properties. This can be expressed as

$$Y^{(m+1)} = \sum_i \left(A_i^{(m)} \right) \frown \mathcal{M} \tag{4}$$

where $\mathcal{M}$ is a vector of the selected molecular properties. This gives the first hidden representation for the molecule. Standard hidden layers which are defined as

$$Y^{(l+1)} = \mathcal{H}^{(l)} \left(Y^{(l)}; W^{(l)} \right) \tag{5}$$

are then applied to this representation, which finally gives the predicted molecular properties. The $\mathcal{H}$ in Eq. (5) represents an arbitrary activation function. In this paper the Leaky ReLU function will be used for $\mathcal{H}^{(l)}$ except in the final layer where the sigmoid function will be used instead. The sigmoid function is selected here since we want the output of our model to be in the range of 0 to 1.

4   Experiments

To demonstrate the improvements that can be achieved by adding more information to the model, we conduct several experiments, each with gradually increasing level of provided molecular information.

4.1   Data

To train and test our proposed model we use the publicly available SIDER dataset[1] presented in [19]. This dataset contains information on molecules, marketed as medicines, and their recorded side effects. The data originates from the SIDER database. The original data consist of 1430 molecules and 5868 different types of side effects. However, in the dataset presented by Wu et al. [19], similar side effects are grouped together, leaving the dataset with 28 groups of side effects. The molecules are stored as SMILES strings [18] and are read into the program and converted to a graph structure using the cheminformatics open source software RDKit [9]. RDKit is also used to extract information about the atoms, bonds and molecules, such as chirality and molecular weight. For a fair comparison between the presented results and the benchmark presented in Wu et al. [19], the dataset is split into the same sets for training, validation, and testing.

4.2   Experimental Design

In order to show that when more information is added to the model presented in Sect. 3 the predictive performance increases, we conduct five different experiments, each using more information about the molecule than the previous experiment. In the most simplistic experiment only the elemental type of the atom is used. In the latter experiments, more information is used and information concerning the bond and the full molecular structure is also incorporated into the model. To this end, the following five set-ups are used:

1. The elemental type of each atom.
2. The elemental type of each atom and its hybridization type.
3. The elemental type of each atom, its hybridization type and the type of each bond.
4. Information concerning the atom, including elemental and hybridization type, chirality, the number of hydrogen atoms the atom binds to, if the atom is in a ring and if that ring is an aromatic ring. Information about the type of bond will also be used.
5. The same information as used in set-up 4, adding information concerning the full molecule. Information such as weight, charge and the number of rotatable bonds in the molecule.

---

[1] https://github.com/deepchem/deepchem/blob/master/examples/sider.

4.3 Implementation

The model described earlier in Sect. 3 is implemented in Theano [15] and with a
network architecture that is as similar as possible as the architecture presented
by Wu et al. [19]. Two convolutional layers, each with 64 neurons, will therefore
bu used. These layers are then followed by a single fully connected layer, with
128 neurons. Between each layer in our model we use a 10% dropout rate [14].
The models are trained by minimizing the cross entropy error. This is achieved
by optimizing the values of all free parameters ($W^{(l)}$ and $V^{(l)}$) using the ADAM
optimization algorithm [7]. Each network is trained for 200 epochs using a batch
size of 20 examples. The performance of the networks are evaluated in the same
way as Wu et al. [19]. To test if there is a significant difference between these
results a one sided t-test is used.

4.4 Experimental Results

We compare the predictive performance of our previously presented model with
the one presented by Wu et al. [19]. The performance of our model on the presented experimental set-ups is measured as the mean AUC-ROC value, averaged
over all trials and all target variables. The mean values of all trials are presented
in Table 1. A graphical representation of the achieved AUC-ROC values, averaged over all target variables, achieved on the training and test data is shown in
Fig. 1. The differences to the result reported by Wu et al. [19] is also shown in
these figures. All experiments except the fourth, significantly outperformed the
results reported by Wu et al. [19] with a p–value < 0.01.

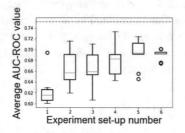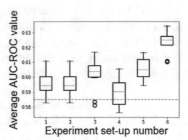

Fig. 1. The distribution of the achieved AUC-ROC values averaged over all target variables for each experimental set-up. The blue dashed line represents the result achieved
by [19]. The left plot is the achieved results on the training data and the right plot is
the achieved results on the test data.

Table 1. The mean AUC-ROC value achieved by the presented model in the described experimental set-ups from the previous subsection. These results are compared with the results in [19] which are shown on the last two rows of the table.

	Experiment					Graph	Random
	1	2	3	4	5	convolution [19]	forest [19]
Train	0.622	0.663	0.666	0.681	0.693	0.751	0.999
Validation	0.554	0.558	0.560	0.586	0.589	0.613	0.670
Test	0.595	0.595	0.602	0.589	0.605	0.585	0.644

5   Discussion and Conclusion

The achieved results presented in Sect. 4.4 were significantly better than those previously obtained on the same dataset with the use of a DCNN. However there are still other models, such as random forest, achieving better results than our model on the given problem. Nonetheless, there are improvements to our model that may change this. The presented model has the advantage of being flexible on what information that can be included and therefore, in contrast to models relying on fingerprints, there is a potential to add more information and hence improve the results. Hence, there should be an opportunity to achieve better results by finding a the optimal set of information to use as well as better network architectures.

Any deep neural network would theoretically perform as good as or better than before when more information is added to the network. The reason for this is that the network can always learn to ignore the additional information. This argumentation follows the same chain of reasoning that He et al. [4] used when arguing for why adding more layers to a DNN should theoretically always improve the performance of the network. However, while it is possible to argue that adding more information to the model always improves the predictive power, this is not always true in practical experiments. If too much irrelevant information is added to the network there is always a risk that the relevant information will vanish in the noise or that the model would be overfitted and thus, it is more difficult to train the model. An example of this is experiment set-up 4 where the AUC-ROC increases for the training data while the AUC-ROC decreases for the test data, as shown in Fig. 1. In Experiment set-up 5, more information about the molecule is added. Much of this information is actually redundant and could be calculated from other features. The molecular weight, which can be calculated from the elemental types of the atoms, is an example of this. However adding such global information at a late stage of the network, facilitates the networks ability to learn other useful features.

Wu et al. [19] recognised the prediction of side effects in the SIDER dataset to be a difficult problem and hypothesised that the reason is that the dataset consisted of biological molecules. To the contrary, we believe that the low average AUC-ROC value is mainly caused by the imbalance in the data for some side effects. This skewness can also, to some extent, explain the big differences between the training and the test error. Our reasoning is further strengthened by that Wu et al. [19] achieved the best result with the random forest algorithm, which is not surprising since the random forest algorithm is more robust to unbalanced classes. Therefore we hypothesise deep learning models for molecular property prediction can be greatly improved by using methods commonly used to handle imbalanced classes. However this is left for further research.

References

1. Dahl, G.E., Jaitly, N., Salakhutdinov, R.: Multi-task neural networks for QSAR predictions. ArXiv e-prints, June 2014
2. Duvenaud, D.K., Maclaurin, D., Iparraguirre, J., Bombarell, R., Hirzel, T., Aspuru-Guzik, A., Adams, R.P.: Convolutional networks on graphs for learning molecular fingerprints. In: Cortes, C., Lawrence, N.D., Lee, D.D., Sugiyama, M., Garnett, R. (eds.) Advances in Neural Information Processing Systems 28, pp. 2224–2232. Curran Associates Inc., Red Hook (2015)
3. Gawehn, E., Hiss, J.A., Schneider, G.: Deep learning in drug discovery. Mol. Inform. **35**(1), 3–14 (2016)
4. He, K., Zhang, X., Ren, S., Sun, J.: Deep residual learning for image recognition. In: Proceedings of the IEEE Conference on Computer Vision and Pattern Recognition, pp. 770–778 (2016)
5. Huuskonen, J., Salo, M., Taskinen, J.: Aqueous solubility prediction of drugs based on molecular topology and neural network modeling. J. Chem. Inf. Comput. Sci. **38**(3), 450–456 (1998)
6. Kearnes, S., McCloskey, K., Berndl, M., Pande, V., Riley, P.: Molecular graph convolutions: moving beyond fingerprints. J. Comput. Aided Mol. Des. **30**(8), 595–608 (2016)
7. Kingma, D.P., Ba, J.: Adam: a method for stochastic optimization. ArXiv e-prints, December 2014
8. Kuhn, M., Letunic, I., Jensen, L.J., Bork, P.: The sider database of drugs and side effects. Nucleic Acids Res. **44**(D1), D1075–D1079 (2016)
9. Landrum, G.: Rdkit: open-source cheminformatics (2006). http://www.rdkit.org. Accessed 3 Apr 2017
10. LeCun, Y., Bengio, Y., Hinton, G.: Deep learning. Nature **521**(7553), 436–444 (2015)
11. Lusci, A., Pollastri, G., Baldi, P.: Deep architectures and deep learning in chemoinformatics: the prediction of aqueous solubility for drug-like molecules. J. Chem. Inf. Model. **53**(7), 1563 (2013)
12. Maas, A.L., Hannun, A.Y., Ng, A.Y.: Rectifier nonlinearities improve neural network acoustic models. In: Proceedings of ICML, vol. 30 (2013)
13. Mayr, A., Klambauer, G., Unterthiner, T., Hochreiter, S.: Deeptox: toxicity prediction using deep learning. Front. Environ. Sci. **3**, 80 (2016)
14. Srivastava, N., Hinton, G.E., Krizhevsky, A., Sutskever, I., Salakhutdinov, R.: Dropout: a simple way to prevent neural networks from overfitting. J. Mach. Learn. Res. **15**(1), 1929–1958 (2014)
15. Theano Development Team: Theano: a Python framework for fast computation of mathematical expressions. arXiv e-prints, abs/1605.02688, May 2016
16. Varnek, A., Baskin, I.: Machine learning methods for property prediction in chemoinformatics: quo vadis? J. Chem. Inf. Model. **52**(6), 1413–1437 (2012)

17. Wallach, I., Dzamba, M., Heifets, A.: AtomNet: a deep convolutional neural network for bioactivity prediction in structure-based drug discovery. ArXiv e-prints, October 2015
18. Weininger, D.: Smiles, a chemical language and information system. 1. Introduction to methodology and encoding rules. In: Proceedings Edinburgh Math. SOC, vol. 17, pp. 1–14 (1970)
19. Wu, Z., Ramsundar, B., Feinberg, E.N., Gomes, J., Geniesse, C., Pappu, A.S., Leswing, K., Pande, V.: MoleculeNet: a benchmark for molecular machine learning. ArXiv e-prints, March 2017

An Analysis of Symmetric Words in Human DNA: Adjacent vs Non-adjacent Word Distances

Carlos A. C. Bastos[1,2]([⊠]), Vera Afreixo[1,3,4], João M. O. S. Rodrigues[1,2], and Armando J. Pinho[1,2]

[1] IEETA-Institute of Electronic Engineering and Informatics of Aveiro, Aveiro, Portugal
[2] Department of Electronics, Telecommunications and Informatics, University of Aveiro, Aveiro, Portugal
cbastos@ua.pt
[3] CIDMA-Center for Research and Development in Mathematics and Applications, Aveiro, Portugal
[4] Department of Mathematics, University of Aveiro, Aveiro, Portugal

Abstract. It is important to develop methods for finding DNA sites with high potential for the formation of hairpin/cruciform structures. In a previous work, we studied the distances between adjacent reversed complement words (symmetric words), and we observed that for some words some distances were favored. In the work presented here, we extended the study to the distance between non-adjacent reversed complement words and we observed strong periodicity in the distance distribution of some words. This may be an indication of potential for the formation of hairpin/cruciform structures.

Keywords: Cruciform · Distance distribution · Genomic word
Reversed complement

1 Introduction

Several genomic studies have focused on the analysis of word counts and word distances. Namely, phylogeny studies [8], alignment-free methods [1,4], CpG detection [6], coding detection [2] and DNA structure analysis [10].

A DNA word analysis based on the distribution of the distances between adjacent symmetric words of length seven was performed [10], and the distributions showed a strong overrepresentation of distances up to 350, a feature that may be associated with the occurrence of hairpin/cruciform structures. However, the cruciform structure can occur between reversed complements that are not necessarily adjacent. The stem and loop lengths of cruciform structures seem to vary over a wide range. According to different authors, the stem length varies between 6 and 100 nucleotides, while loop lengths may range from 0 to 2000 nucleotides [5,7,11]. The aim of this work is to analyse the occurrence of adjacent and non-adjacent symmetric words along the sequence.

© Springer Nature Switzerland AG 2019
F. Fdez-Riverola et al. (Eds.): PACBB 2018, AISC 803, pp. 80–87, 2019.
https://doi.org/10.1007/978-3-319-98702-6_10

2   Methods

The human genome is the subject of this study and the main purpose is to explore the human DNA structure. Specifically, we want to explore structures beyond the well-known repetition structures. Thus, we used pre-masked sequences available from the UCSC Genome Browser (http://genome.ucsc.edu) downloads page. These files contain the GRCh38 assembly sequences, with repeats reported by RepeatMasker [9] and Tandem Repeats Finder [3] masked with Ns.

Consider the alphabet $\mathcal{A} = \{A, C, G, T\}$ and let w be a symbolic sequence (word) defined in $\mathcal{A}^k$, where k is the length of w. The pair composed by one word, w, and the corresponding reversed complement word, w', is called a symmetric word pair. For example, (ACT, AGT) is a symmetric word pair.

For a given word length k, we compute the frequency distributions of distances between occurrences of each word and the adjacent reversed complement word $f_{w,w'}$. We also compute the frequency distributions of distances between occurrences of each word and all succeeding reversed complements, $f_{w,w'...w'}$. We compare both distance distributions using the well-known Kullback-Leibler divergence (KL).

For example, consider the following sequence:

$$ACTGGAA\overline{AGT}AAG\overline{AGT}\underline{ACT}TTGT\underline{ACT}GGG\overline{AGT}TTGT$$

For word $w = ACT$ we have only two valid distances between adjacent reversed complement words (7 and 6), but we have five distances between adjacent and non adjacent reversed complement words (7, 14, 30, 13, 6).

We analyse distances up to 4000 nucleotides, but, if an N symbol is found, the search for w' is stopped. To avoid the direct word dependencies, we exclude distances shorter than k. Motivated by previous work, computational limitations and the stem length of possible cruciform structures, we study $k = 7$.

In order to analyse abundant words in the DNA sequence, we exclude symmetric word pairs with relative occurrence frequency lower than $1/4^7$.

3   Results and Discussion

Table 1 shows the 10 words with the greatest divergence between $f_{w,w'}$ and $f_{w,w'...w'}$. Note that there are no CG pairs in the composition of the words and that there is no obvious pattern in words composition.

By visual comparison of the distance distributions of several words, we found distinct patterns of divergence between adjacent and non-adjacent words. Figures 1 and 2 show two of those divergence patterns. For word $w = TGTCACC$, the distribution of $f_{w,w'}$ shows a single strong peak at distance 28 (see Fig. 1 top), whereas the $f_{w,w'...w'}$ distribution displays a periodic pattern of peaks at distances of $28+48i$, with $i = 0, 1, 2, \ldots$ (see Fig. 1 bottom). For word $w = TGCATGC$, the divergence between $f_{w,w'}$ and $f_{w,w'...w'}$ is also high. The $f_{w,w'}$ distribution has a single strong peak at distance 324, which is weakened

Table 1. The ten words with greatest divergence between $f_{w,w'}$ and $f_{w,w'...w'}$. For each word, the Kullback-Leibler divergence and the nucleotide composition are shown.

Word	KL	n_A	n_C	n_G	n_A
TGTGCAC	1.175	1	2	2	2
TGGGCCC	1.141	0	3	3	1
GGAGCTC	1.054	1	2	3	1
TGGGTAA	1.053	2	0	3	2
TTACCCA	1.033	2	3	0	2
TGTCCAC	0.961	1	3	1	2
GGGCCCA	0.941	1	3	3	0
AGAATTC	0.918	3	1	1	2
TGTCACC	0.911	1	3	1	2
TGCATGC	0.905	1	2	2	2

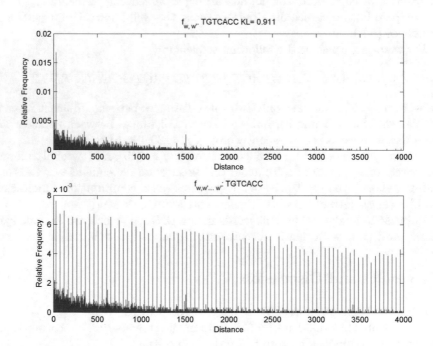

Fig. 1. $f_{w,w'}$ and $f_{w,w'...w'}$ *TGTCACC*.

in $f_{w,w'...w'}$, but the extra peaks in the $f_{w,w'...w'}$ distribution do not introduce additional regularity (see Fig. 2).

With our exploratory analysis by visual inspection, we found an interesting regular pattern in words with high divergence between $f_{w,w'}$ and $f_{w,w'...w'}$: words with a few peaks in $f_{w,w'}$ and with several nearly periodic peaks in $f_{w,w'...w'}$

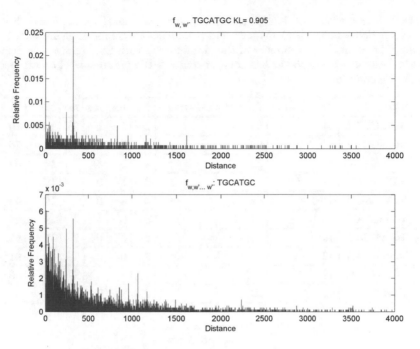

Fig. 2. $f_{w,w'}$ and $f_{w,w'...w'}$ $TGCATGC$.

(see for example Fig. 1). To find words that have this divergence pattern, we implemented the following algorithm:

– compute
  - $m = \max\{f_{w,w'}\}$
  - $n_1 = \#\{d \in 1,\ldots,4000 : f_{w,w'}(d) > cm\}$
  - $n_2 = \#\{d \in 1,\ldots,4000 : f_{w,w'...w'}(d) > cm\}$, with $c \in]0;1[$

– select the words ($w \in \mathcal{A}^k$) for which $n_1 \leq 2$ and $n_2 - n_1 > 2$

For $c = 0.4$, the algorithm selected 34 words. Table 2 shows the 10 words with exactly periodic peaks in $f_{w,w'...w'}$. It is interesting to note that the peak period of 6 out of 10 distributions has a value of 84. The most frequent spacing between peaks of $f_{w,w'...w'}$, for each word, is not longer than 102.

In order to characterize the chromosomal distribution of the peak distances (d_p) we searched, in each chromosome, the positions where the word w appears at a distance d_p before w' and we counted the number of occurrences in each chromosome. Table 2 contains the chromosome with the highest percentage of occurrence of the first peak distance. It is evident that the pattern of distance distributions studied here reflects a local behaviour that occurs mainly in a single chromosome. Moreover, all of the words of Table 2 with the same peak period occurred in the same chromosome.

Table 2. Words with a peak in $f_{w,w'}$ and regular peaks in $f_{w,w'...w'}$. For each word, the peak period, the first peak distance and the nucleotide composition are shown. The column Chr* contains the number of the chromosome with the highest occurrence of the first peak distance and the last column contains the percentage of occurrences at the Chr* chromosome.

Word	Peak period	First peak distance	n_A	n_C	n_G	n_T	Chr*	%
AAGCTTT	84	83	2	1	1	3	19	70
AGGCCTT	84	83	1	2	2	2	19	76
AGTGTGG	84	52	1	0	4	2	19	34
ATTCATA	84	21	3	1	0	3	19	54
CCACACT	84	32	2	4	0	1	19	43
TATGAAT	84	63	3	0	1	3	19	55
TCACCAT	44	36	2	3	0	2	7	91
TGGGTAA	42	13	2	0	3	2	11	91
TGTCACC	48	28	1	3	1	2	3	86
TTACCCA	42	29	2	3	0	2	11	91

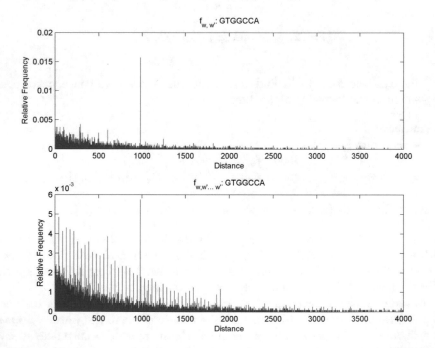

Fig. 3. $f_{w,w'}$ and $f_{w,w'...w'}$ $GTGGCCA$.

Figures 3 and 4 show examples of the distance distributions found by the previous algorithm for words $w = GTGGCCA$ and $w = TCACCAT$. Both words have the expected regular pattern of peaks in $f_{w,w'...w'}$. Figure 3 presents an interesting pattern with a very strong peak at a distance of 980. Figure 4 also

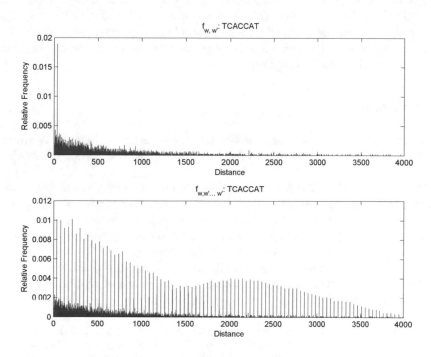

Fig. 4. $f_{w,w'}$ and $f_{w,w'...w'}$ $TCACCAT$.

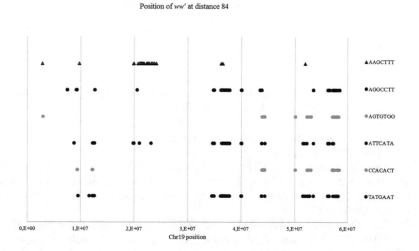

Fig. 5. Positions of the first peak of ww' distance for six words with peak period of 84 (in chromosome 19).

shows a regular pattern of peaks in $f_{w,w'...w'}$, with a heavier tail, resembling a mixture of two distributions.

Figure 5 shows the positions, in chromosome 19, of the first peak distance of the words in Table 2 with peak period of 84. These positions seem to form three groups of words: group 1 - *AGGCCTT*, *ATTCATA* and *TATGAAT*; group 2 - *AGTGTGG* and *CCACACT*; and group 3 - *AAGCTTT*. Both group 1 and group 2 contain each other a reverse complement pair which might indicate that the words w and w' form a regular pattern of occurrence in the regions of chromosome 19 shown in the figure.

4   Conclusions

We believe that strong overrepresentation of some distances between symmetric words is a feature that may be associated with the occurrence of cruciform structures.

Our analysis identified a set of words with unusual distribution of distances to the corresponding reversed complements. Since we use masked sequences, the observed regularities are not due to the known repeated structures.

The regular periodic pattern of the distance between reversed complements occurs mostly at some regions of a single chromosome.

We expect that this analysis contributes to clarify the possible association between the features of distances between symmetric words and the occurrence of cruciform structures.

Acknowledgment. This work was supported by FEDER ("Programa Operacional Fatores de Competitividade" COMPETE) and FCT ("Fundação para a Ciência e a Tecnologia"), within the projects UID/MAT/04106/2013 to CIDMA (Center for Research and Development in Mathematics and Applications) and UID/CEC/00127/2013 to IEETA (Institute of Electronics and Informatics Engineering of Aveiro).

References

1. Afreixo, V., Bastos, C.A.C., Pinho, A.J., Garcia, S.P., Ferreira, P.J.S.G.: Genome analysis with inter-nucleotide distances. Bioinformatics **25**(23), 3064–3070 (2009)
2. Bastos, C.A.C., Afreixo, V., Garcia, S.P., Pinho, A.J.: Inter-stop symbol distances for the identification of coding regions. J. Integr. Bioinform. **10**(3), 31–39 (2013)
3. Benson, G.: Tandem repeats finder: a program to analyze DNA sequences. Nucleic Acids Res. **27**(2), 573 (1999)
4. Bernard, G., Chan, C.X., Chan, Y.-B., Chua, X.-Y., Cong, Y., Hogan, J.M., Maetschke, S.R., Ragan, M.A.: Alignment-free inference of hierarchical and reticulate phylogenomic relationships. Brief. Bioinform., bbx067 (2017). https://doi.org/10.1093/bib/bbx067
5. Cer, R.Z., Bruce, K.H., Mudunuri, U.S., Yi, M., Volfovsky, N., Luke, B.T., Bacolla, A., Collins, J.R., Stephens, R.M.: Non-B DB: a database of predicted non-B DNA-forming motifs in mammalian genomes. Nucleic Acids Res. **39**(suppl. 1), D383–D391 (2010)

6. Hackenberg, M., Previti, C., Luque-Escamilla, P.L., Carpena, P., Martínez-Aroza, J., Oliver, J.L.: CpGcluster: a distance-based algorithm for CpG-island detection. BMC Bioinform. **7**(1), 446 (2006)
7. Kolb, J., Chuzhanova, N.A., Högel, J., Vasquez, K.M., Cooper, D.N., Bacolla, A., Kehrer-Sawatzki, H.: Cruciform-forming inverted repeats appear to have mediated many of the microinversions that distinguish the human and chimpanzee genomes. Chromosome Res. **17**(4), 469–483 (2009)
8. Sims, G.E., Kim, S.-H.: Whole-genome phylogeny of Escherichia coli/Shigella group by feature frequency profiles (FFPs). Proc. Nat. Acad. Sci. **108**(20), 8329–8334 (2011)
9. Smit, A.F.A., Hubley, R., Green, P.: Repeatmasker (1996)
10. Tavares, A.H.M.P., Pinho, A.J., Silva, R.M., Rodrigues, J.M.O.S., Bastos, C.A.C., Ferreira, P.J.S.G., Afreixo, V.: DNA word analysis based on the distribution of the distances between symmetric words. Sci. Rep. **7**(1), 728 (2017)
11. Wang, Y., Leung, F.C.C.: Long inverted repeats in eukaryotic genomes: recombinogenic motifs determine genomic plasticity. FEBS Lett. **580**(5), 1277–1284 (2006)

A Bioinformatics Protocol for Quickly Creating Large-Scale Phylogenetic Trees

Hugo López-Fernández[1,2,3,4,5(✉)], Pedro Duque[4,5,6],
Sílvia Henriques[4,5], Noé Vázquez[1,2], Florentino Fdez-Riverola[1,2,3],
Cristina P. Vieira[4,5], Miguel Reboiro-Jato[1,2,3], and Jorge Vieira[4,5]

[1] ESEI - Escuela Superior de Ingeniería Informática,
Edificio Politécnico, Universidad de Vigo,
Campus Universitario As Lagoas s/n, 32004 Ourense, Spain
{hlfernandez, riverola, mrjato}@uvigo.es,
nvazquezg@gmail.com
[2] Centro de Investigaciones Biomédicas
(Centro Singular de Investigación de Galicia), Vigo, Spain
[3] SING Research Group, Galicia Sur Health Research Institute (IIS Galicia Sur).
SERGAS-UVIGO, Vigo, Spain
[4] Instituto de Investigação e Inovação em Saúde (I3S), Universidade do Porto,
Rua Alfredo Allen, 208, 4200-135 Porto, Portugal
{pedro.duque, silvia.henriques}@i3s.up.pt,
{cgvieira, jbvieira}@ibmc.up.pt
[5] Instituto de Biologia Molecular e Celular (IBMC),
Rua Alfredo Allen, 208, 4200-135 Porto, Portugal
[6] Faculdade de Ciências, Universidade do Porto,
Rua do Campo Alegre 1021/1055, 4169-007 Porto, Portugal

Abstract. The large scale genome datasets that are now available can provide unprecedented insight into the evolution of genes and gene families. Nevertheless, handling and transforming such datasets into the desired format for downstream analyses is often a difficult and time-consuming task for researchers without a background in informatics. Here, we present a simple and fast protocol for data preparation and high quality phylogenetic tree inferences using simple to install cross-platform software applications with rich graphical interfaces. To illustrate its potential, this protocol was used to provide insight into the evolution of *GULO* gene in animals, a gene that encodes the enzyme responsible for the last step of vitamin C synthesis in this group of organisms. We find that *GULO* is always a single copy gene in all animal groups with the exception of Echinodermata. Surprisingly, we find potentially functional *GULO* genes in several Prostotomian groups such as Molluscs, Priapulida and Arachnida. To our knowledge, this is the first time a putative functional *GULO* gene is reported in Protostomians. All previously reported *GULO* gene losses were easily identified using the presented protocol.

Keywords: Large scale analyses · GULO · Animals

© Springer Nature Switzerland AG 2019
F. Fdez-Riverola et al. (Eds.): PACBB 2018, AISC 803, pp. 88–96, 2019.
https://doi.org/10.1007/978-3-319-98702-6_11

1 Introduction

Genome sequence data is accumulating at an explosive pace, and although many available genomes are non-annotated, there are still hundreds of annotated genomes that can be used to address the evolution of genes and gene families at an unprecedented scale, using a phylogenetic approach. Getting the data into the adequate format for running the phylogenetic analyses, and in a format as close as possible to what is intended to appear in the final picture, is, however, not a trivial issue for researchers without an informatics background. Therefore, here, we report on a tested protocol for quickly getting the data into the desired format, which can be used by any researcher regardless of their informatics skills. By using objective criteria, data analyses are also highly reproducible. High quality phylogenies, using hundreds of sequences, can be obtained in a couple of days, being most of this time actually spent downloading the data and running the phylogenetic analysis, and not preparing the data.

We provide an example of the usefulness of such protocol, by making inferences on the evolution, in animals, of the GULO gene, which is involved in the last step of the vitamin C (ascorbic acid) synthesis pathway. Ascorbic acid is an antioxidant vitamin needed for the formation of collagen, for healthy teeth, gums and blood vessels. Moreover, it improves iron absorption and resistance to infection, and is essential for brain development [1]. In adults, insufficient vitamin C in the diet leads to the lethal deficiency disease scurvy. Despite its importance, it has been reported that anthropoid primates, guinea pigs, some bat species, some Passeriformes bird species, and teleost fishes, have lost the capacity to synthesize vitamin C due to mutations in the *GULO* gene (reviewed in [2]). The independent loss of *GULO* in several animal lineages has been argued to be a neutral event for animals that have a vitamin C rich diet [2]. Extensive analyses on non-vertebrate animal species have, however, been lacking. Here we use all available annotated animal genome sequences to estimate the number of times that this gene has been lost and duplicated.

2 Materials and Methods

The protocol for quick recovery of the coding sequences (CDSs) of interest and reformatting of FASTA files is summarized in Fig. 1. CDSs were downloaded from NCBI by querying for "Animals" under the "Assembly" option[1]. Both GenBank and RefSeq annotations were downloaded, since the data in the two databases overlaps only partially. All operations regarding reformatting of files and CDS filtering were performed using SEDA[2], which within a single step, applies each given operation to a large number of files (458 files in our case). An example of the reformatting operations is given in Fig. 2. When dealing with a large number of files, care must be taken with possible unintended contaminations. Therefore, using SEDA's "NCBI Rename" option we added as a prefix to each file name, information on the species name, common

---

[1] https://www.ncbi.nlm.nih.gov/assembly/.

[2] http://www.sing-group.org/seda/.

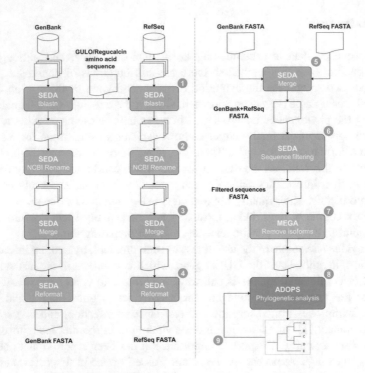

Fig. 1. Schematic flow diagram of the proposed protocol.

A. Original GenBank sequence

```
lcl|AKHW03001036.1_cds_KYO43973.1_5236 [locus_tag=Y1Q_0005872]
[protein=L-gulonolactone oxidase-like] [protein_id=KYO43973.1]
[location=join(310796..310798,313412..313513,313822..313947,31511
7..315211,318930..319020,322717..322902,339211..339318,339509..33
9585,342102..342251,343194..343358,344699..344789,345872..345988,
346102..346176)] [gbkey=CDS]
ATGGTTCACGGCCATGGAGGCGTCCAGTTCCAGAACTGGGCCAAGACCTATGGC
```

B. GenBank sequence after NCBI Rename and reformat header

```
>Alligator_mississippiensis_(American_alligator)_Alligatoridae_KY
O43973.1
ATGGCAGCCAAACGGAATAATGTCACCACAGACGTGCCGGAATCATCTGTGGCGCCGGC
```

Fig. 2. FASTA file sample before and after being processed by SEDA.

name, and kingdom to which the species belongs to. Although we searched for "Animals", annotations were retrieved for one bacterial and one viral species. These two FASTA files were removed from downstream analyses. We also downloaded from RefSeq the genome annotation of five fungi species to be used as an outgroup in phylogenetic analyses. At this stage, GenBank and RefSeq FASTA files were processed independently. Since animal GenBank and RefSeq FASTA files amount to 6.3 and 26.4 GB, respectively, the next step was to try to reduce as much as possible the size of the files being used, and thus, using SEDA, we performed a *tblastn* search on each FASTA file separately, using as query the *Mus musculus* (the house mouse) L-gulonolactone oxidase (GULO) protein sequence available in ENSEMBL (ENSMUSP00000060912). An expect value of 0.05 was used and we imposed no limit for the number of BLAST hits to retrieve. It should be noted that the name of the resulting file is the name of the query sequence plus the name of the original FASTA file used as subject. Next, we used SEDA (NCBI Rename option) to add as a prefix to the header of each of the retrieved sequences, the name of the species, common name, and the family name to which the species belongs to. For both GenBank and RefSeq data, using the "Merge" option, files were merged into a single file. Line breaks were removed from the resulting file using the "Reformat file" option, and the headers were reformatted using the "Rename header" option in order to keep only the species name, common name, family name, and accession number. The two files were again merged using the "Merge" option, and using the "Pattern filtering" option, sequences with ambiguous positions (Ns) were removed, as well as those not showing the typical amino acid HWXK motif [3]. GULO belongs to the vanillyl-alcohol oxidase (VAO) flavoproteins family that includes GULO homologs, such as L-galactono-1,4-lactone dehydrogenases (GalDH) and D-arabinono-1,4-lactone oxidases (AraO), which catalyze the last step of L-ascorbate or D-erythorbate synthesis, respectively. The HWXK motif is conserved amongst GULO, GalDH and AraO sequences from mammals, plants and yeasts, respectively, and thus all GULO sequences are expected to show such a pattern. Sequences retrieved by BLAST that do not show such a pattern are highly unlikely to be functional GULO sequences. In animals there are no GalDH and AraO genes. Identical nucleotide sequences were then removed using the "Remove redundant sequences" option. The list of merged headers was exported to check if different species had identical nucleotide sequences. None was found. Sequences that are non-multiple of three, that do not have a valid start codon, and that have in frame stop codons were then removed using the SEDA "Filtering" option. The *M. musculus* sequence was reallocated to the beginning of the file using the "Reallocate reference sequences" option, and the "Filtering" option used to remove sequences with a size difference larger than 10% relative to the first (i.e., *M. musculus*) sequence. The "Sort" option was used to alphabetically order the sequences according to the species name.

A neighbour-joining phylogeny was obtained using MEGA [4], in order to identify and remove CDS isoforms. This is the only time-consuming step that cannot be easily automated. The final FASTA format was then used to obtain a Bayesian phylogenetic tree using MrBayes, as implemented in ADOPS, after alignment with Muscle [5; see references therein]. We have chosen ADOPS because it offers a convenient pipeline for obtaining Bayesian phylogenetic trees using a set of unaligned coding sequences in FASTA format, and the use of different alignment algorithms. The resulting tree, in Nexus format, was converted to Newick using the Format Conversion Website[3]. The resulting tree was then rooted using MEGA software, and the final picture lightly edited.

---

[3] http://phylogeny.lirmm.fr/phylo_cgi/data_converter.cgi.

3 Results and Discussion

The detection of one single species showing a GULO with all expected features (similarity with reference sequence, presence of the typical amino acid pattern, expected size, and expected position in the phylogenetic tree) is enough to state that *GULO* has not been lost in the lineage to which the species belong to. Nevertheless, a *GULO* sequence with the expected features could be missing due to the incompleteness of the genome sequence, or failure to annotate the gene. Therefore, in order to consider the hypothesis that *GULO* is missing in a given lineage, we require a minimum of three species where a *GULO* CDS with all expected features is not detected. Special attention is given to the case where sequences have all expected features, except the expected size, since it could be a case of miss-annotation of a functional *GULO* gene, or miss-annotation due to the presence of in-frame stop codons, in which case *GULO* is a pseudogene. Figure 3 summarizes the findings regarding *GULO* presence/absence in animal lineages, while Supplementary Fig. 1 (available at http://dx.doi.org/10.5281/zenodo.1163119) shows the phylogenetic relationship of all 118 *GULO* functional sequences that we have identified. The nine sequences that have passed the first three filters but that are not placed in the expected position in the tree are unlikely to represent the *GULO* gene. They could be the result of bacterial contamination of the samples used for genome sequencing, or represent genes unrelated to *GULO* that also have a flavin adenine dinucleotide (FAD) domain. We did not attempt to equilibrate the number of species in the different branches of the tree, because we wanted to address whether analyzing all available data could produce new biological insight.

The *GULO* gene is present in non-bilaterian species from the Cnidaria group (Fig. 3 and Supplementary Fig. 1), as reported by Wheeler [6]. We identified a potentially functional *GULO* gene sequence in three out of five species of the Anthozoa class. Although no putative functional *GULO* gene sequence was found for both Mixozoa and Hydrozoa a single species was analysed, and thus no conclusions are taken. The bilaterian group can be divided into Prostotomians and Deuterostomians. We detect a putative functional *GULO* gene in four Prostotomian species, namely two Mollusc, one Arachnida and one Priapulida species. Within Molluscs, a putative functional *GULO* is present in Gastropods (present in two out of three species analysed), but absent in Bivalvia (four species analysed). Since a single Cephalopoda species was analysed, it is not safe to conclude yet that *GULO* is not present in this group of species, since GULO may have not been annotated in this genome, or be missing because it was not possible to assemble this genome region. Within the Arachnida class, we failed to detect a putative functional *GULO* gene in eight species of the Acari subclass, although in two cases a partial *GULO* sequence has been detected. One of the two Acari partial sequences may be the result of wrong gene annotation and may be a functional *GULO* sequence. Nevertheless, a putative functional *GULO* sequence was detected in one (*Parasteatoda tepidariorum*) of the two species of Araneae. A *tblastn* search using as query the *M. musculus* GULO protein against the genome of the Araneae species that did not make it until the end of the protocol (*Stegodyphus mimosarum*), shows that there are no obvious in frame stop codons that could suggest that this species has a *GULO* pseudogene, although we need

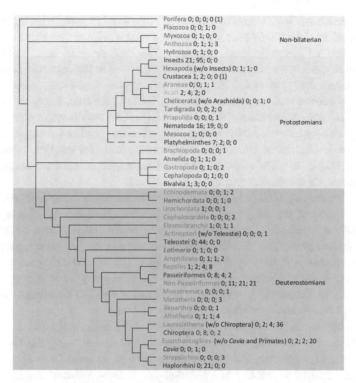

Fig. 3. Summary of the findings regarding the presence of putative functional *GULO* genes in different animal lineages. In green, red, blue, violet and orange are, respectively, the lineages where a likely functional *GULO* has been detected, lineages where a functional *GULO* gene has not been detected, lineages for which there is little data and thus no firm conclusions can be made, lineages showing a mix of species with a functional and non-functional *GULO* gene, and lineages where most species show a non-functional *GULO* gene but where at least one species may have a functional *GULO* gene. The first three numbers after the names indicate the number of species that were removed from the analysis because: (1) no sequence was found with significant homology with the query sequence when performing BLAST; (2) sequences did not show the typical GULO amino acid pattern or showed ambiguous positions; (3) sequences do not show a valid start codon, are non-multiple of three, have in frame stop codons, or have a size difference larger than 10% relative to the reference sequence. The last number indicates the number of species present in the final tree. Numbers in brackets indicate species that made it to the final tree but whose sequences fall in the wrong place and are thus unlikely to be represent *GULO* sequences. Broken lines show uncertain relationships. Taxonomic relationships are depicted as in the Tree of life web project (http://tolweb.org) and in Helgen [7].

to bear in mind that the similarity does not extend over the entire sequence. More detailed analyses are needed to elucidate how many Araneae species have a putative functional *GULO* sequence, and understand if Acari species do have a functional *GULO* gene. One Hexapoda species, one Chelicerata and one Annelida species where a partial *GULO* sequence has been annotated, likely have a functional *GULO* gene, while Tardigrades likely do not.

The detection of *GULO* gene sequences in Protostomian species is surprising given that we failed to detect *GULO* in three large Protostomian data sets: 117 insect, 25 Nematode, and 10 Platyhelminthes genomes. The Protostomian *GULO* sequences do not seem to be the result of contamination, since they are in the expected position in the tree. Moreover, they are not *GULO* gene paralogous sequences, since the *Ciona intestinalis* sequence that, as expected, is more basal in the tree, has been identified as being a *GULO* gene sequence by Wheeler *et al.* [6]. Our results imply that *GULO* was lost multiple times independently within Protostomians, and not a single time in the Protostomian lineage, after the split of the Protostomian and the Deuterostomian lineages. Analysing all available data was clearly advantageous, since we could have been misled if a few Protostomian species were analysed.

In Deuterostomes, *GULO* is readily identified in many animal groups with the notable exception of teleost (bony) fishes, as previously reported ([2] and references therein). Although, we have analysed 44 fish genomes in none of them we were able to detect the *GULO* gene. This was the expected result since it has been reported that no teleost fish species is able to synthesize vitamin C [2]. No *GULO* gene was detected either in *Latimeria chalumnae* (coelacanth), a fish species more closely related to tetrapods than to Teleostei. As expected, we detect a putative functional *GULO* gene sequence in two Elasmobranch species (sharks and rays).

Within birds, the Galliformes, Gruiformes, Pelecaniformes, and Passeriformes orders were further analysed by performing a *tblastn* search using the *GULO* mouse sequence as query, because, for each of these groups, at least three species showed features that could be suggestive of a pseudogene, namely, a size difference larger than 10% relative to the reference CDS, absence of a start codon, sequences non-multiple of three, or having in frame stop codons. This analysis, revealed that *GULO* gene is not well annotated in the Galliformes and Pelecaniformes genomes, but since there are no in-frame stop codons, and all exons could be found, they likely have a functional *GULO* gene. In the case of Gruiformes, the GULO sequence is complete with the exception of the first methionine. Nevertheless, the second codon is preceded by the nucleotides AG, which is compatible with the location of one intron with canonical GT-AG splice sites at this location. If this is the case, then it will be difficult to find the previous exon in the absence of mRNA data. As previously reported, within Passeriformes, several species do not present a functional *GULO* gene, although three species (*Acanthisitta chloris*, *Lonchura striata* and *Ficedula albicollis*) are not well annotated, and seem to have a functional *GULO* gene. Three squamata (Reptiles) partial GULO sequences are likely the result of a bad annotation, and likely have a functional *GULO* gene as well.

Anthropoid primates, guinea pigs, and some bat species have been reported to have lost the capacity to synthesize vitamin C as well. In agreement with these findings, although all 21 primate genomes showed at least one hit when performing the *tblastn* step, none showed the typical GULO amino acid pattern. Moreover, in only two (*Rousettus aegyptiacus* and *Rhinolophus sinicus*) out of ten bat species, a putative functional *GULO* gene is found, as previously reported for other *Rousettus* and *Rhinolophus* species [8]. It is noteworthy to mention that, although a *GULO* gene is annotated in the guinea pig (*Cavia porcellus*) genome, the protein encoded by this putative functional gene is half the size of that found in other species, and thus it is a pseudogene. This sequence is not present in the phylogenetic tree presented in Supplementary Fig. 1 because it was eliminated in the relative size sequence removal step, showing how useful this option can be.

The phylogenetic tree here obtained also shows that, in animals, *GULO* is a single copy gene in all species analysed with the exception of *Acanthaster planci* (Echinodermata). Therefore, during evolution, *GULO* duplicates were not retained after the two rounds of closely spaced auto-tetraploidization events (usually referred to as 1R and 2R) that occurred early in vertebrate evolution [9, 10]. These are thought to have occurred during chordate evolution, after the split of the urochordate and cephalochordate lineages and before the radiation of gnathostomes. It is likely that *GULO* gene had already been lost in the common ancestor of all Teleosti fish, and thus the whole-genome duplication event that occurred at that time [11] is not relevant. Moreover, even in recent allotetraploid species such as *Xenopus leavis*, a single *GULO* gene is found [12].

4   Conclusions

The protocol here described can be used to study the evolution of any gene or gene family, and requires no previous bioinformatics expertise or special informatics equipment. All software tools here used can be used in any platform (Windows, MacOS and Linux), either because they are truly cross-platform or because they are available as virtual disk installations as well. Most steps do not require much human intervention, with the exception of the removal of CDS isoforms. By using such protocol, we were able to show that: (a) *GULO* is present in both non-bilaterian and bilaterian animals; (b) *GULO* gene copies have never been retained after the whole genome duplication events that occurred in different animal lineages, and thus is always a single copy gene with the exception of the starfish; (c) *GULO* is detected for the first time in two Protostomian species; (d) *GULO* has been lost independently several times, both within Protostomes and Deuterostomes. In the future, we intend to modify this protocol in order to integrate information from non-annotated genomes as well. Such protocol would allow the use of almost 1300 animal genomes rather than the 454 annotated genomes here used. It would also, allow the removal of the time-consuming CDS isoform removal step, and possibly to reduce the problems that are cause by the wrong CDS annotation, when performing phylogenetic analyses.

Acknowledgements. This article is a result of the project Norte-01-0145-FEDER-000008 - Porto Neurosciences and Neurologic Disease Research Initiative at I3S, supported by Norte Portugal Regional Operational Programme (NORTE 2020), under the PORTUGAL 2020 Partnership Agreement, through the European Regional Development Fund (FEDER). Financial support from the Xunta de Galicia (Centro singular de investigación de Galicia accreditation 2016-2019) and the European Union (European Regional Development Fund - ERDF), is gratefully acknowledged. H. López-Fernández is supported by a post-doctoral fellowship from Xunta de Galicia (ED481B 2016/068-0).

References

1. Patananan, A.N., Budenholzer, L.M., Pedraza, M.E., Torres, E.R., Adler, L.N., Clarke, S.G.: The invertebrate Caenorhabditis elegans biosynthesizes ascorbate. Arch. Biochem. Biophys. **569**, 32–44 (2015)
2. Drouin, G., Godin, J.-R., Page, B.: The genetics of vitamin C loss in vertebrates. Curr. Genomics **12**, 371–378 (2011)
3. Leferink, N.G.H., Jose, M.D.F., van den Berg, W.A.M., van Berkel, W.J.H.: Functional assignment of Glu386 and Arg388 in the active site of l-galactono-γ-lactone dehydrogenase. FEBS Lett. **583**, 3199–3203 (2009)
4. Kumar, S., Stecher, G., Tamura, K.: MEGA7: molecular evolutionary genetics analysis version 7.0 for bigger datasets. Mol. Biol. Evol. **33**, 1870–1874 (2016)
5. Reboiro-Jato, D., Reboiro-Jato, M., Fdez-Riverola, F., Vieira, C.P., Fonseca, N.A., Vieira, J.: ADOPS–Automatic Detection Of Positively Selected Sites. J. Integr. Bioinform. **9**, 200 (2012)
6. Wheeler, G., Ishikawa, T., Pornsaksit, V., Smirnoff, N.: Evolution of alternative biosynthetic pathways for vitamin C following plastid acquisition in photosynthetic eukaryotes. eLife. **4** (2015)
7. Helgen, K.M.: The mammal family tree. Science **334**, 458–459 (2011)
8. Cui, J., Yuan, X., Wang, L., Jones, G., Zhang, S.: Recent loss of vitamin C biosynthesis ability in bats. PLoS ONE **6**, e27114 (2011)
9. Putnam, N.H., Butts, T., Ferrier, D.E.K., Furlong, R.F., Hellsten, U., Kawashima, T., Robinson-Rechavi, M., Shoguchi, E., Terry, A., Yu, J.-K., Benito-Gutiérrez, E., Dubchak, I., Garcia-Fernàndez, J., Gibson-Brown, J.J., Grigoriev, I.V., Horton, A.C., de Jong, P.J., Jurka, J., Kapitonov, V.V., Kohara, Y., Kuroki, Y., Lindquist, E., Lucas, S., Osoegawa, K., Pennacchio, L.A., Salamov, A.A., Satou, Y., Sauka-Spengler, T., Schmutz, J., Shin-I, T., Toyoda, A., Bronner-Fraser, M., Fujiyama, A., Holland, L.Z., Holland, P.W.H., Satoh, N., Rokhsar, D.S.: The amphioxus genome and the evolution of the chordate karyotype. Nature **453**, 1064–1071 (2008)
10. Dehal, P., Boore, J.L.: Two rounds of whole genome duplication in the ancestral vertebrate. PLoS Biol. **3**, e314 (2005)
11. Taylor, J.S.: Genome duplication, a trait shared by 22,000 species of ray-finned fish. Genome Res. **13**, 382–390 (2003)
12. Session, A.M., Uno, Y., Kwon, T., Chapman, J.A., Toyoda, A., Takahashi, S., Fukui, A., Hikosaka, A., Suzuki, A., Kondo, M., van Heeringen, S.J., Quigley, I., Heinz, S., Ogino, H., Ochi, H., Hellsten, U., Lyons, J.B., Simakov, O., Putnam, N., Stites, J., Kuroki, Y., Tanaka, T., Michiue, T., Watanabe, M., Bogdanovic, O., Lister, R., Georgiou, G., Paranjpe, S.S., van Kruijsbergen, I., Shu, S., Carlson, J., Kinoshita, T., Ohta, Y., Mawaribuchi, S., Jenkins, J., Grimwood, J., Schmutz, J., Mitros, T., Mozaffari, S.V., Suzuki, Y., Haramoto, Y., Yamamoto, T.S., Takagi, C., Heald, R., Miller, K., Haudenschild, C., Kitzman, J., Nakayama, T., Izutsu, Y., Robert, J., Fortriede, J., Burns, K., Lotay, V., Karimi, K., Yasuoka, Y., Dichmann, D.S., Flajnik, M.F., Houston, D.W., Shendure, J., DuPasquier, L., Vize, P.D., Zorn, A.M., Ito, M., Marcotte, E.M., Wallingford, J.B., Ito, Y., Asashima, M., Ueno, N., Matsuda, Y., Veenstra, G.J.C., Fujiyama, A., Harland, R.M., Taira, M., Rokhsar, D.S.: Genome evolution in the allotetraploid frog Xenopus laevis. Nature **538**, 336–343 (2016)

Deep Learning Analysis of Binding Behavior of Virus Displayed Peptides to AuNPs

Haebom Lee[1](✉), Jun Jo[1], Yong Oh Lee[1], Korkmaz Zirpel Nuriye[1](✉), and Leon Abelmann[1,2]

[1] KIST Europe, Saarbrücken, Germany
haebom.lee@gmail.com, N.Korkmaz@kist-europe.de
[2] University of Twente, Enschede, The Netherlands

Abstract. Filamentous fd viruses have been used as biotemplates to develop nano sized carriers for biomedical applications. Genetically modified fd viruses with enhanced gold binding properties have been previously obtained by displaying gold binding peptides on viral coat proteins. In order to generate a stable colloidal system of dispersed viruses decorated with AuNPs avoiding aggregation, the underlying binding mechanism of AuNP-peptide interaction should be explored. In this paper, we therefore propose a macro scale self-assembly experiment using 3D printed models of AuNP and the virus to extend our understanding of Au binding process. Moreover, we present our image analysis algorithm which combines image processing techniques and deep learning to automatically examine the coupling state of the particles.

Keywords: Deep learning · Genetic engineering · Phage display Gold

1 Introduction

Non-toxic and cost effective stable carrier systems are important for drug delivery and cancer therapy applications. Although gold nanoparticles (AuNPs) can be synthesized chemically using reducing agents with defined geometries and sizes, the functionalization of particles in order to obtain targetable agents still remains as a challenge. Until now AuNPs have been functionalized with antibodies, biotin molecules or cell targeting moieties using chemical reaction processes [1]. An alternative way of functionalization avoiding chemical modifications is to use specific Au binding moieties like short peptides which would selectively bind Au. Au binding peptides can be tagged with cell targeting sequences and the resulting "AuNP-peptide-targeting unit" complex can be used for drug delivery, cancer therapy or imaging studies [2].

After identification of peptides, it is crucial to test their functionalities first. We have previously generated genetically engineered fd viruses expressing various short peptide sequences and investigated their Au binding affinities using biochemical and microscopical analyses [3,4]. Genetically modified viruses showed

© Springer Nature Switzerland AG 2019
F. Fdez-Riverola et al. (Eds.): PACBB 2018, AISC 803, pp. 97–104, 2019.
https://doi.org/10.1007/978-3-319-98702-6_12

enhanced Au binding properties. These viruses were tested as potential biotemplates for nanowire synthesis taking the advantage of improved Au binding characteristics. However, to obtain single dispersed virus templated nanowires was challenging due to the aggregation problem which we face after the metallization process. In order to control the aggregation of viral particles during Au binding process, we need to understand how they interact with the particles and which factors may affect the binding behavior.

Until now, computer simulation analysis like molecular dynamics (MD) studies have been performed to investigate the underlying binding mechanism of Au binding peptides to Au surfaces. Hnilova et al. [5] have studied the peptide-metal (Au, Pt, Pd) surface binding using different surface facets (111, 100) considering various peptide properties such as stability, length and flexibility. Yu et al. [6] conducted MD analyses on Au surface binding process of a previously identified Au binding peptide A3 (AYSSGAPPMPPF). In their study, computer simulation analyses were conducted without any supporting experimental data after building up 3D conformational structures of defined peptides. For the simulation studies, they started with a non-binding state in solution and they modeled the Au-peptide interactions while approaching the Au surface. Binding energies were calculated for each amino acid residue as the energy difference before and after the binding process. Chiappini et al. [7] have published a numerical study on phase behavior of fd viruses interacting with DNA functionalized AuNPs. Due to the interaction between complementary strands of DNA fragments, an aggregated colloidal system of viruses and AuNPs was obtained by experimental analyses. In their numerical models employing Monte Carlo simulations, the authors assumed that the colloidal system is a mixture of hard spheres of AuNPs and rods of viruses. Interaction energies between virus and AuNPs were calculated at different temperatures in order to model the nucleation and aggregation behavior of the binary system. In general, however, MD simulations are time consuming when a large number of objects is involved, numerical artifacts accumulate over time, and generally MD algorithms minimize the potential energy only and do not include an entropy term.

Recently, we have created an experimental setup to study self-assembly processes on the macroscale [8]. The system consists of 3D printed polymer objects with embedded permanent magnets in side an upward flow of water that provides levitation as well as turbulence. Compared to MD simulation, the evaluation time is independent on the number of objects. The attempt frequency is in the order of 10 Hz, which is significantly faster than MD simulations [9]. On the other hand only dipolar forces can be modeled. Also turbulence is not exactly identical to thermal agitation in the sense that there is a dependency on length scale and direction. Macroscale self-assembly studies however do provide fast and complementary information to MD simulations, especially in the first stages of analysis.

In contrast to MD simulations, the position of objects in the macroscale studies need to be observed by cameras. Analysis of the self-assembly process can be done manually, but often is a tedious job that takes longer than the self-assembly experiment itself. Automated image recognition can be used for simple situations. In this paper, we propose the use of big data analysis using deep neural networks.

2 Self-Assembly Experiment

To investigate and simulate the AuNP-virus interaction, which is induced by Au binding peptides displayed on virus coat proteins (Fig. 1a), a 3D self-assembly experiment is designed using scaled-up solid models of virus and Au particles employing magnetic intermolecular forces.

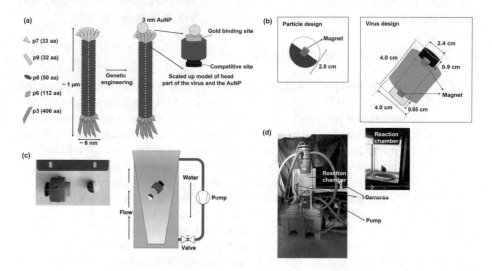

Fig. 1. (a) Schematic drawing of an fd filamentous virus composed of a single stranded circular DNA enclosed in a protein cage. Head part of the virus is genetically modified to display Au binding peptides. The part is considered as a cylindrical object in our 3D scaled-up model. (b) Yellow part stands for the Au binding site, whereas black part indicates the competitive binding site with less binding energy. (c) 3D printed models of viral template and the AuNP. Solid objects are inserted into (d) the self-assembly reactor chamber and recorded using two cameras under turbulent flow.

Figure 1b shows the simplified designs of virus and Au particle. The extended cylindrical part in yellow represents the head part of the fd virus where Au binding peptide is displayed. The extruding cylindrical part in black acts as the competitive binding site. The center to center distance between the magnet in the black part and the AuNP mimicking spherical particle is arranged to be 19.0 mm when they are combined, which is 0.5 mm longer than the distance between the magnet of the yellow cylindrical part and the gold particle. As a result, we expect to have stronger binding of AuNP mimicking sphere on the yellow part.

Particles are designed using a free 3D modeling tool [10] and 3D printed using ABS polymer (Fig. 1c). Axially magnetized cylindrical NdFeB magnets (5 mm in length and 4 mm in diameter, Supermagnete, grade N42, Webcraft GmbH), are glued inside each hole of the cylindrical subunits and assembled together.

Following the particle design and 3D printing, we organize a self-assembly trial experiment in a flow reactor chamber as illustrated in Fig. 1c–d. The macroscopic self-assembly reactor is based on that of Hageman et al. [8]. A water pump forms a turbulent flow against the gravitational force inside the conical reaction chamber and is introducing turbulence as the source of disturbing energy. We use a MAXI.2 40T pump moving water through four inlets into the cone shaped inset keeping the objects in the center of the observation area. Two calibrated, synchronized cameras (Mako G-131, Allied Vision) are perpendicularly positioned at the front and right sides of the reaction chamber for image recording.

3   Binding Pattern Analysis

Image processing techniques for finding meaningful features are widely exploited in bioinformatics area [11]. Our analysis algorithm also employs an image processing step to remove an uninformative background and extract a meaningful region of interest (ROI) from the captured images. We first calculate the difference between a test image and a clean background image, and apply a threshold to obtain a binary mask for the ROI. Then we extract red and yellow pixels from the ROI, as the analysis should recognize the case as positive only when the red part of the gold particle and the yellow part of the virus are combined (Fig. 2). We also count the number of connected regions in the extracted images, in order to provide additional data to our CNN classifier. The CNN classifier also receives the combined image of red and yellow channels of ROI as an input. The entire image processing steps and their intermediate results are depicted in Fig. 3.

(a) Positive   (b) Negative   (c) Negative   (d) Negative   (e) Negative

Fig. 2. Examples of analysis results for various input images: (a) two particles are combined at the Au binding site; (b) connected, but at the competitive binding site; (c) particles are separated; (d) and (e) one or no particle is captured in the image.

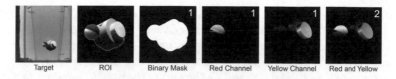

Fig. 3. The pipeline of our image processing procedure. From a target image, region of interest is extracted and further investigated through its binary image, red channel, yellow channel, and a combined image of red and yellow channels. Numbers on images indicate the number of connected components in each image. The rightmost combined image and four metrics are the input of the CNN.

Images acquired from the self-assembly experiments should be labeled before being used as a training data for our CNN. Instead of conducting arduous manual classification, we exploit synthetic images that can be generated from a 3D game engine. Utilizing synthetic data to train a neural network is gaining popularity in modern deep learning applications [12] as it provides flexibility and control at the same time and can generate ground truth data with a lower cost.

Our aim is to achieve maximum similarity between the real and the synthetic data. For this reason, the 3D Openscad object design data is exploited again to construct a basic scene of the simulation. Then we applied arbitrary motions to the virtual particles. Furthermore, we locate the virtual lamps and cameras cautiously, to imitate our self-assembly experiment environment. Figure 4 shows the rendered synthetic images and corresponding image processing results. We generate synthetic images for various cases and use them to train and test our CNN classifier.

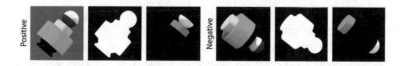

Fig. 4. Images generated from a 3D game engine and their subsequent image processing results.

3.1   Deep Learning

We extract several metrics from the input images and deliver them as an additional input to the network. For a pair of input images (front and right), we calculate four metrics from each image: the number of connected components in ROI's BW, red channel, yellow channel, and red+yellow channel images.

Figure 5 shows the overall architecture of our CNN. The network mainly consists of three convolution blocks and two fully-connected blocks, as well as flattening and concatenating layers that bridge the two types of blocks. Each convolution block is composed of a convolution layer with ReLU activation, followed by a max pooling layer, and a dropout layer. The first convolution layer filters the input of size $1280 \times 1024 \times 2$ using 16 kernels of size 10×10 with a horizontal stride of 10 pixels and a vertical stride of 8 pixels. The result image of size $128 \times 128 \times 16$ is then shrunk to $64 \times 64 \times 16$ in the following max pooling layer. The second convolution layer filters it using 32 kernels of size 3×3 with a stride of 2 pixels. Through a similar process, the input of the third convolution layer becomes of size $16 \times 16 \times 32$ and at the end of the block, the input is flattened to a vector of length 1024 from $4 \times 4 \times 64$. The vector is then concatenated with the second input, a vector of eight metrics, and delivered to a fully-connected layer of 512 neurons. Finally, the vector of length 512 goes through a sigmoid function to generate the output.

Fig. 5. Overall architecture of our CNN classifier. Layers with same color have the same setting unless otherwise stated.

4   Experiment

In the self-assembly experiment, we set an upward flow in the reactor chamber of 6.5(1) cm/s using a valve setting with an equivalent energy estimated to be 10(2) µJ. To measure the binding energy between the virus model and the gold particle, we exploit a dipole model as illustrated in Fig. 6 since the distance between the magnets is large compared to their size (see Fig. 1). When the sphere is connected to the yellow side of the cylinder, the distance between the magnets is 19.3(2) mm and the energy is –94(3) µJ. When it is connected to the thick black side, the distance is 19.8(2) mm and the energy consequently higher, –87(3) µJ. The energy difference between the two states is 7(6) µJ.

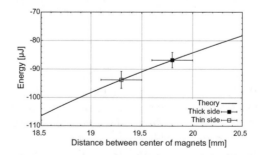

Fig. 6. Calculated energy between two cylindrical magnets of diameter 4 mm and height 5 mm as a function of the distance between the magnets. The two points indicated are the distances between the magnets when the sphere is on the thick (black) side or thin (yellow) side of the main cylinder.

We conducted the self-assembly experiment for four hours and recorded images of size 1280 × 1024 at 1 fps, leading to over 14,000 pairs of front and right images. We labeled 3,200 randomly chosen pairs manually and left the remaining unlabeled. In total, 725 positive real pairs, 2,475 negative real pairs, 12,000 positive synthetic pairs, and 12,000 negative synthetic pairs are utilized in our experiment. The classification accuracies of our classifier in different settings are presented in Table 1. Specifically, we trained our CNN model in three different ways: training only with synthetic data, training only with real data, and training with a combined set of synthetic and real data. Then we evaluated accuracy of the classifier for the training set and for the 3,200 labeled real data. Note that the presented accuracies are the mean accuracy over 10-fold cross validation except the case when only the synthetic data is used as a training set.

Our algorithm reaches an accuracy of 93.2% when it is trained with labeled real images. It justifies our utilization of synthetic data, as it recorded 88.7% accuracy when only the real data are used as a training set. Our classifier also shows promising results with synthetic images, reaching an accuracy of 86%. These accuracies are much better than an image processing based naive classifier, which examines images using the number of connected components in the red+yellow channel. It does not achieve an accuracy above 77%. Furthermore, our classifier, trained with real images, estimates the fraction of positives in the remaining 10,800 unlabeled real image pairs to be 22.2%, which is very close to the 22.7% of the labeled set.

Table 1. Accuracies of our classifier over training data and 3,200 real image pairs. Each column represents a different composition of training data, when we utilized only 24,000 synthetic image pairs, only 3,200 real image pairs, and a combined set of entire real and synthetic image pairs, respectively.

	Synthetic	Real	Synthetic + Real
Training	99.2%	97.4%	99.5%
Real Data	85.9%	88.7%	93.2%

5   Conclusion and Future Work

Considering increasing significance of particle-peptide based hybrid materials for a broad range of applications including biomedical and catalysis studies, it is important to understand the binding mechanism of peptides. To this end, we devised and conducted macro scale self-assembly experiments with 3D printed objects with embedded magnets in a turbulent flow, simulating a genetically modified virus displaying a gold binding peptide and a gold particle.

Images captured during the self-assembly experiment are then analyzed using a classifier which combines image processing technique and convolutional neural network. We also introduced how we utilized synthetic data to train the classifier in an efficient way. Various measurements over the classifier showed its capability as a reliable combined state analysis algorithm, by achieving 93.2% accuracy which is 16.4 p.p. better than a naive classifier.

In the future, we want to conduct further self-assembly experiments with increased number of Au particles. Moreover, we hope to extend our experimental approach in order to investigate effect of AuNP and virus concentration, peptide display position, number of displayed peptides, and temperature on the self-assembly process. As a consequence, we expect more sophisticated synthetic data generation and more flexible classifier in our future work.

Acknowledgement. This work was supported by KIST Europe Institutional Program [Project No. 11807].

References

1. Giljohann, D.A., Seferos, D.S., Daniel, W.L., Massich, M.D., Patel, P.C., Mirkin, C.A.: Gold nanoparticles for biology and medicine. Angew. Chem. Int. Ed. **49**(19), 3280–3294 (2010)
2. Kumar, A., et al.: Gold nanoparticles functionalized with therapeutic and targeted peptides for cancer treatment. Biomaterials **33**(4), 1180–1189 (2012)
3. Korkmaz, N.: Recombinant bacteriophages as gold binding bio-templates. Colloids Surf. B Biointerfaces **112**, 219–228 (2013)
4. Zirpel, N.K., Arslan, T., Lee, H.: Engineering filamentous bacteriophages for enhanced gold binding and metallization properties. J. Colloid Interface Sci. **454**, 80–88 (2015)
5. Hnilova, M., et al.: Effect of molecular conformations on the adsorption behavior of gold-binding peptides. Langmuir **24**(21), 12440–12445 (2008)
6. Jing, Y., Becker, M.L., Carri, G.A.: The influence of amino acid sequence and functionality on the binding process of peptides onto gold surfaces. Langmuir **28**(2), 1408–1417 (2011)
7. Chiappini, M., Eiser, E., Sciortino, F.: Phase behaviour in complementary DNA-coated gold nanoparticles and fd-viruses mixtures: a numerical study. Eur. Phys. J. E **40**(1), 7 (2017)
8. Hageman, T.A.G., Löthman, P.A., Dirnberger, M., Elwenspoek, M.C., Manz, A., Abelmann, L.: Macroscopic equivalence for microscopic motion in a turbulence driven three-dimensional self-assembly reactor. J. Appl. Phys. **123**(2), 024901 (2018)
9. Biyikli, E., To, A.C.: Multiresolution molecular mechanics: Implementation and efficiency. J. Comput. Phys. **328**, 27–45 (2017)
10. Openscad. http://www.openscad.org/
11. Teixeira-Castro, A., et al.: An image processing application for quantification of protein aggregates in caenorhabditis elegans. In: 5th International Conference on Practical Applications of Computational Biology & Bioinformatics (PACBB 2011), pp. 31–38. Springer (2011)
12. Richter, S.R., Vineet, V., Roth, S., Koltun, V.: Playing for data: ground truth from computer games. In: European Conference on Computer Vision, pp. 102–118. Springer (2016)

Compression of Amino Acid Sequences

Diogo Pratas$^{(\boxtimes)}$, Morteza Hosseini, and Armando J. Pinho

IEETA/DETI, University of Aveiro, Aveiro, Portugal
{pratas,seyedmorteza,ap}@ua.pt

Abstract. Amino acid sequences are known to be very hard to compress. In this paper, we propose a lossless compressor for efficient compression of amino acid sequences (AC). The compressor uses a cooperation between multiple context and substitutional tolerant context models. The cooperation between models is balanced with weights that benefit the models with better performance, according to a forgetting function specific for each model. We have shown consistently better compression results than other approaches, using low computational resources. The compressor implementation is freely available, under license GPLv3, at https://github.com/pratas/ac.

Keywords: Lossless compression · Amino acid sequences
Protein entropy

1 Introduction

An amino acid sequence is a succession of letters, usually with 20 outcomes, that indicate the order and nature of amino acids within a protein. Each amino acid may have several corresponding DNA codon, taken from the 4-symbol DNA (or RNA) alphabet. The information flow seems to be only unidirectional: from DNA to RNA to amino acids to proteins.

In Table 1, we show the amino acids representation, possible DNA codon(s) and its distribution percentage over two large protein databases. As it can be seen, each amino acid may have multiple codon representations. Additionally, it can be represented with more than 20 outcomes, mostly given the uncertainty provided from the DNA or RNA sequencing process.

The compression of amino acid sequences is known to be a very hard challenge [1]. Some authors even propose that these sequences were approximately uncompressible, given the marginal compression gains provided by their proposed algorithm, CP [2]. These results were also supported with the fact that some compression algorithms need more bits to represent the sequence than those in the original sequences.

For example, the theoretical value to store an amino acid, assuming that the cardinality is 20, is $\log_2 20 \approx 4.322$. If we use Gzip to compress an amino acid sequence, most of the times it will need more than its theoretical value. This happens mainly because Gzip has a short memory model.

© Springer Nature Switzerland AG 2019
F. Fdez-Riverola et al. (Eds.): PACBB 2018, AISC 803, pp. 105–113, 2019.
https://doi.org/10.1007/978-3-319-98702-6_13

Table 1. Amino Acids representation, possible DNA codon(s) and its distribution percentage over two large protein databases. Bold values stand for amino acids that have a representability higher than 5% in at least one of the databases. The underlined names represent the start and stop codons. The databases, DB1 (uniprot_sprot.fasta.gz) and DB2 (uniprot_trembl.fasta.gz), have been downloaded from ftp://ftp.uniprot.org/pub/databases/uniprot/current_release/knowledgebase/complete/.

Letter	Amino acid	DB1%	DB2%	Possible DNA codon(s)
A	Alanine	**8.26**	**9.10**	GCT, GCC, GCA, GCG
C	Cysteine	1.37	1.21	TGT, TGC
D	Aspartic acid	**5.46**	**5.45**	GAT, GAC
E	Glutamic acid	**6.73**	**6.16**	GAA, GAG
F	Phenylalanine	3.86	3.92	TTT, TTC
G	Glycine	**7.08**	**7.26**	GGT, GGC, GGA, GGG
H	Histidine	2.27	2.19	CAT, CAC
I	Isoleucine	**5.93**	**5.70**	ATT, ATC, ATA
K	Lysine	**5.82**	4.99	AAA, AAG
L	Leucine	**9.65**	**9.87**	TTA, TTG, CTT, CTC, CTA, CTG
M	<u>Start,</u> Methionine	2.41	2.38	ATG
N	Asparagine	4.06	3.88	AAT, AAC
P	Proline	4.72	4.85	CCT, CCC, CCA, CCG
Q	Glutamine	3.93	3.79	CAA, CAG
R	Arginine	**5.53**	**5.71**	CGT, CGC, CGA, CGG, AGA, AGG
S	Serine	**6.61**	**6.69**	TCT, TCC, TCA, TCG, AGT, AGC
T	Threonine	**5.35**	**5.57**	ACT, ACC, ACA, ACG
V	Valine	**6.86**	**6.88**	GTT, GTC, GTA, GTG
W	Tryptophan	1.09	1.29	TGG
Y	Tyrosine	2.92	2.93	TAT, TAC
*	<u>Stop</u>	?	?	TTA, TAG, TGA
B	D or N	0.00	**5.03**	GAC, GAT, AAC, AAT
Z	E or Q	0.00	0.00	GAA, GAG, CAA, CAG
X	<any>	0.00	0.04	<any>

Generally, the difficulty to compress amino acids sequences is given by the lack of redundancy between approximate regions. To prove this, Benedetto *et al.* shown the relevance of medium-range correlations in the compression of several amino acids sequences, proposing an estimation known as Model3 [3].

For exploring medium and long-range correlations, namely through the identification of duplicates/repeats, Hategan and Tabus [4] proposed ProtComp, mainly modeling approximate repeats and optimal building of the substitution probability matrix. The prediction by partial matching (PPM) scheme [5] was also used in amino acid sequences, although it was developed much earlier. The same happened with the context-tree weighting (CTW) [6], although Matsumoto *et al.* proposed a CTW-based algorithm, predicting the probability of a character by weighting the importance of short and long contexts, using the occurrence of approximate repeats or palindromes in the contexts [7].

Cao *et al.* proposed XM [8], with substantial improvements over previous algorithms. The XM method uses a combination of three types of models: (1) order-2 context models; (2) order-1 context models (typically using information from the 512 previous symbols); (3) the copy expert, that considers the next symbol as part of a copied region from a particular offset.

Other algorithms have also been proposed, namely BWT/SCP [9] and CAD [10]. The BWT/SCP [9] algorithm is based on the Burrows-Wheeler transform (BWT) [11] and the Sorted Common Prefix (SCP) [12]. On the other hand, CAD is based on an adaptive dictionary. Although these algorithms improved substantially the compression ratio, they have been used in different datasets, more prone to redundancies [3]. For a deeper review of amino acid sequence compression, see [13].

Besides storage purposes, the compression of amino acid sequences is very important to predict and uncover structure [1,14], namely through fusions and duplications in proteins, that may be linked with new functionalities [15]. More examples are protein classification [16] and domain identification [17].

In this paper, we propose a new method to compress amino acid sequences (AC). We extend a well-known dataset and, finally, we compare it against the state-of-the-art compressors.

2   Method

The method consists in a cooperation between context and substitutional tolerant context models of several depths that are weighted, according to its performance, given a specific forgetting function for each model. We describe its foundations in the following subsections.

2.1   Context Models

Consider an information source that generates symbols from Θ and that it has already generated the sequence of n symbols $x^n = x_1 x_2 \ldots x_n$, $x_i \in \Theta$ A subsequence of x^n, from position i to j, is denoted as x_i^j.

A Context model, m, is a statistical model assuming the Markov property, that assigns probability estimates to the symbols of Θ, according to a conditioning context computed over a finite and fixed number, k, of past outcomes (order-k) [18]. Usually $m = k$, where k is the context size. However, we may have different m with the same k, such as a context model and a substitutional tolerant context model with the same context size.

The probabilities are estimated using the parameter α. We address the information content estimation process under the form

$$P_m(s|x_{i-k}^{i-1}) = \frac{N(s|x_{i-k}^{i-1}) + \alpha}{N(x_{i-k}^{i-1}) + \alpha|\Theta|}, \tag{1}$$

where $N(s|x_{i-k}^{i-1})$ represents the number of times that, in the past, the information source generated symbol s having x_{i-k}^{i-1} as the conditioning context and where $N(x_{i-k}^{i-1})$ is the total number of events that has occurred so far in association with context x_{i-k}^{i-1}. The parameter α allows balancing between the maximum likelihood estimator and a uniform distribution. For deeper orders, α should be generally lower than one.

2.2   Substitutional Tolerant Context Models

A substitutional tolerant context models (STCM) [19,20] is a probabilistic-algorithmic context model. It assigns probabilities according to a conditioning context that considers the last symbol, from the sequence to occur, as the most probable, given the occurrences stored in the memory instead of the true occurring symbol.

For a symbol $s \in \Theta$, the estimator of a STCM is given by

$$P_m(s|x'^{i-1}_{i-k}) = \frac{N(s|x'^{i-1}_{i-k}) + \alpha}{N(x'^{i-1}_{i-k}) + \alpha|\Theta|}, \tag{2}$$

where function N accounts for the memory counts regarding the model and x' is a copy of x, edited according to

$$x'_i = \operatorname*{argmax}_{\forall s \in \Theta} P_m(s|x'^{i-1}_{i-k}). \tag{3}$$

An STCM, besides being probabilistic, is also algorithmic, namely because they can be switched on or off given its performance, according to a threshold, t, defined before the computation. The t enables or disables the model, according to the number of times that the context has been seen, given k hits or fails that are constantly stored in memory in a cache array.

2.3   Cooperation

The probability of the next symbol, x_{n+1}, is given by

$$P(x_{n+1}) = \sum_{m \in \mathcal{M}} P_m(x_{n+1}|x_{n-k+1}^n)\, w_{m,n}, \tag{4}$$

$P_m(x_{n+1}|x_{n-m+1}^n)$ where is the probability assigned to the next symbol by a context or substitutional tolerant context model, and where $w_{m,n}$ denote the corresponding weighting factors, with

$$w_{m,n} \propto (w_{m,n-1})^{\gamma_m} P_m(x_n|x_{n-k}^{n-1}), \tag{5}$$

where $\gamma_m \in [0, 1)$ acts as a forgetting factor for each context model. The weights are constrained to

$$\sum_{m \in \mathcal{M}} w_{m,n} = 1. \qquad (6)$$

Notice that the fundamental difference in the weights is given by setting each γ_m. Generally, we have found by exhaustive search that models with lower m are more prone to use lower γ_m (typically, below 0.9), while higher m, is associated with higher γ_m (near 0.95). This means that, in this mixture type, the forgetting intensity should be lower for more complex models.

2.4   Implementation

The tool (AC) has been written in C language and it is available, under the GPLv3 license, at http://github.com/pratas/ac. We have set, by default, seven running modes (characterized by parameterizations) that have been estimated given the authors experience. Automatic optimization models might be achieved in future works, consequently improving the compression results.

3   Dataset

Given the low number of proteomes available to compare the existing amino acid sequence compressors in the Protein corpus http://data-compression.info/ Corpora/ProteinCorpus, we have extended the dataset adding five more proteomes. The whole dataset, containing the nine proteomes, is available at http:// sweet.ua.pt/pratas/datasets/AminoAcidsCorpus.zip. Its characteristics can be seen in Table 2.

Table 2. Dataset containing 9 proteomes. The length is given by the size of each amino acid sequence. The datasets HI, MJ, HS and SC are from http://data-compression. info/Corpora/ProteinCorpus, while the rest have been downloaded from biotechnology repositories.

File	Species	Length	Cardinality	Domain	Kingdom
HI	*Haemophilus influenzae*	509,519	20	Bacteria	Eubacteria
MJ	*Methanococcus jannaschii*	448,779	20	Archaea	Euryarchaeota
HS	*Homo sapiens*	3,295,751	19	Eukaryote	Animalia
SC	*Saccharomyces cerevisiae*	2,900,352	20	Eukaryote	Fungi
EC	*Escherichia coli*	1,308,765	21	Bacteria	Eubacteria
EP	*Enterococcus phage*	4,184	20	Virus	Virus
BT	*Bos taurus*	12,845,466	24	Eukaryote	Animalia
LC	*Lactobacillus casei*	809,301	20	Bacteria	Eubacteria
SA	*Staphylococcus aureus*	796,785	20	Bacteria	Eubacteria

We have included amino acid sequences with different cardinalities, lengths and from different Domains and Kingdoms, which enables a more fair and wider comparison regarding the compression skills of each tool.

4 Results

We used the dataset described in Sect. 3 to benchmark open source compressors. We also included some results (for LZ-CTW, Model3, ProtComp, CP and XM) already computed in [21].

Table 3 describes the benchmark. As it can be seen, for the dataset composed with 9 proteomes, AC achieves consistently better compression results. The exception is the smallest file (with ≈4k). In the worst case, AC used 3.6 GB of RAM, spending less than 9 min for compressing the complete dataset. All the compressions performed by AC are able to be losslessly decoded.

Table 3. Average number of bits needed to represent each amino acid for several methods in four datasets (defined in Table 2). The average compression of individual files are given in bits peer symbol. The AC-α is the preliminary AC version (without STCM and mixture improvement). The AC-β is the AC version with STCMs but without mixture improvement. The AC is the final version (with STCMs and mixture improvement). The "AC -l" identifies the parameterization setup used. The symbol "*" stands for results unable to obtain by lack of program availability. The "Model3" includes only estimated results (not real compression).

Methods	HI	MJ	HS	SC	EC	EP	BT	LC	SA
Gzip	4.671	4.587	4.605	4.639	4.679	4.686	4.521	4.655	4.646
Bzip2	4.324	4.269	4.255	4.299	4.324	4.485	4.254	4.302	4.300
7za	4.293	4.206	4.028	4.130	4.267	4.592	3.212	4.273	4.258
Lzma	4.240	4.147	4.026	4.121	4.257	4.422	3.288	4.249	4.228
Lzma -9	4.238	4.141	4.021	4.029	4.229	4.422	3.208	4.188	4.197
LZ-CTW	4.117	4.027	4.005	3.951	*	*	*	*	*
Model3	**4.100**	4.000	3.930	3.950	*	*	*	*	*
ProtComp	4.108	4.008	3.824	3.938	*	*	*	*	*
CP	4.143	4.051	4.112	4.146	*	*	*	*	*
XM	4.102	4.000	3.786	3.885	*	*	*	*	*
paq8h -8	4.118	4.015	3.922	3.957	4.092	4.328	3.170	4.091	4.080
paq8l -8	4.104	3.999	3.901	3.942	4.077	**4.300**	3.144	4.076	4.061
AC-α	4.123	4.029	3.975	3.976	4.107	4.342	3.236	4.103	4.119
AC-β	4.113	4.008	3.810	3.891	4.050	4.336	3.078	4.067	4.072
AC	**4.100**	**3.997**	**3.785**	**3.876**	**4.037**	4.323	**3.055**	**4.055**	**4.056**
AC -l	2	4	7	5	5	1	6	2	5

The XM approach seems to reach very competitive results (in four files) comparatively to AC. However, these results are based on [21], since we could not access to the protein code class (not an open source algorithm).

The paq8h (http://mattmahoney.net/dc/) has also a good performance, although it needed approximately 5x more time than AC, while AC uses 2x more RAM than paq8h.

The paq8l (http://mattmahoney.net/dc/) has even better compression results than paq8h, and competitive results regarding AC, needing even fewer bits to represent the sequence EP. However it needed approximately 8x more time than AC, while AC uses 2x more RAM than paq8l.

The computational time to run the AC in each sequence from the dataset is less than two minutes, with exception to the BT (13 MB) that needed 3.5 min. The AC compressor ran in a single CPU core at 2.13 GHz, without SSD, with a maximum RAM value below 3.7 GB. For decreasing the RAM, some higher-k models might be discarded, having a small precision payoff.

The substitutional tolerant models (STCM) have been successfully implemented in DNA sequences [19,20]. As we can see in Table 3, they also improve the compression of amino acid sequences (AC-α without STCMs and AC-β with STCMs). Additionally, the mixture improvement, with a forgetting function specific for each model, together with the context models and STCMs, give consistently the best results.

We will maintain the results and add more compressors (http://sweet.ua.pt/pratas/compression.html) in a quest to find the best algorithm that, given an affordable computational resource cost, is able to efficiently describe amino acid sequences. These are very important to predict functionalities, as well as for comparative results and promote the development of intelligent algorithms.

5   Conclusions

We have presented a lossless compressor for efficient compression of amino acid sequences (AC). The compressor uses a cooperation between multiple context and substitutional tolerant context models. The cooperation between models is balanced with weights that benefit the models with better performance, according to a forgetting function specific for each model.

We have extended the size of a very used dataset, from 4 to 9 proteomes, and used it to evaluate the performance of the AC. We have shown that AC achieves consistently better compression results than other compressors, with the exception of the smallest sequence. AC used low computational resources, especially computational time.

Acknowledgments. This work was partially funded by the European Union Seventh Framework Programme (FP7/2007–2013) under grant agreement No. 305444 "RD-Connect: An integrated platform connecting registries, biobanks and clinical bioinformatics for rare disease research". It was also funded by FEDER (Programa Operacional

Factores de Competitividade - COMPETE) and by National Funds through the FCT - Foundation for Science and Technology, in the context of the UID/CEC/00127/2013 and PTCD/EEI-SII/6608/2014.

References

1. Nalbantoglu, Ö.U., Russell, D.J., Sayood, K.: Data compression concepts and algorithms and their applications to bioinformatics. Entropy **12**(1), 34–52 (2009)
2. Nevill-Manning, C.G., Witten, I.H.: Protein is incompressible. In: Data Compression Conference, pp. 257–266 (1999)
3. Benedetto, D., Caglioti, E., Chica, C.: Compressing proteomes: the relevance of medium range correlations. EURASIP J. Bioinf. Syst. Biol. **2007**, 5 (2007)
4. Hategan, A., Tabus, I.: Protein is compressible. In: Proceedings of the 6th Nordic Signal Processing Symposium, NORSIG-2004, Espoo, Finland, pp. 192–195, June 2004
5. Cleary, J.G., Witten, I.H.: Data compression using adaptive coding and partial string matching. IEEE Trans. Commun. **32**(4), 396–402 (1984)
6. Willems, F.M.J., Shtarkov, Y.M., Tjalkens, T.J.: The context-tree weighting method: basic principles. IEEE Trans. Inf. Theor. **41**(3), 653–664 (1995)
7. Matsumoto, T., Sadakane, K., Imai, H.: Biological sequence compression algorithms. In: Dunker, A.K., Konagaya, A., Miyano, S., Takagi, T. (eds.) Genome Informatics 2000: Proceedings of the 11th Workshop, Tokyo, Japan, pp. 43–52 (2000)
8. Cao, M.D., Dix, T.I., Allison, L., Mears, C.: A simple statistical algorithm for biological sequence compression. In: Proceedings of the Data Compression Conference, DCC 2007, Snowbird, Utah, pp. 43–52, March 2007
9. Adjeroh, D., Nan, F.: On compressibility of protein sequences. In: Proceedings of Data Compression Conference, DCC 2006. IEEE (2006). 10 p
10. Nag, A., Karforma, S.: Adaptive dictionary-based compression of protein sequences. Int. J. Educ. Manag. Eng. **5**, 1–6 (2017)
11. Ferragina, P., Manzini, G.: Burrows-Wheeler transform. In: Kao, M.Y. (ed.) Encyclopedia of Algorithms, pp. 1–99. Springer, Boston (2008)
12. Adjeroh, D., Feng, J.: The SCP and compressed domain analysis of biological sequences. In: Proceedings of the 2003 IEEE Bioinformatics Conference, CSB 2003, pp. 587–592. IEEE (2003)
13. Hosseini, M., Pratas, D., Pinho, A.J.: A survey on data compression methods for biological sequences. Information **7**(4), 56 (2016)
14. Korber, B., Farber, R.M., Wolpert, D.H., Lapedes, A.S.: Covariation of mutations in the V3 loop of human immunodeficiency virus type 1 envelope protein: an information theoretic analysis. Proc. Natl. Acad. Sci. **90**(15), 7176–7180 (1993)
15. Hayashida, M., Ruan, P., Akutsu, T.: Proteome compression via protein domain compositions. Methods **67**(3), 380–385 (2014)
16. Pelta, D.A., Gonzalez, J.R., Krasnogor, N.: Protein structure comparison through fuzzy contact maps and the universal similarity metric. In: EUSFLAT Conference, pp. 1124–1129 (2005)
17. Rocha, J., Rosselló, F., Segura, J.: Compression ratios based on the Universal Similarity Metric still yield protein distances far from CATH distances. arXiv preprint q-bio/0603007 (2006)
18. Sayood, K.: Introduction to Data Compression, 3rd edn. Morgan Kaufmann, San Francisco (2006)

19. Pratas, D., Pinho, A.J., Ferreira, P.J.S.G.: Efficient compression of genomic sequences. In: Proceedings of the Data Compression Conference, DCC 2016, Snowbird, Utah, pp. 231–240, March 2016
20. Pratas, D., Hosseini, M., Pinho, A.J.: Substitutional tolerant Markov models for relative compression of DNA sequences. In: 11th International Conference on Practical Applications of Computational Biology and Bioinformatics, pp. 265–272. Springer (2017)
21. Diribi, W., Raimond, K.: Comparison of protein corpuses. Int. J. Innov. Manag. Technol. **3**(3), 281 (2012)

NET-ASAR: A Tool for DNA Sequence Search Based on Data Compression

Manuel Gaspar, Diogo Pratas$^{(\boxtimes)}$, and Armando J. Pinho

IEETA, University of Aveiro, Aveiro, Portugal
{manuel.gaspar,pratas,ap}@ua.pt

Abstract. The great increase in the amount of sequenced DNA has created a problem: the storage of the sequences. As such, data compression techniques, designed specifically to compress genetic information, is an important area of research and development. Likewise, the ability to search similar DNA sequences in relation to a larger sequence, such as a chromosome, has a really important role in the study of organisms and the possible connection between different species. This paper proposes NET-ASAR, a tool for DNA sequence search, based on data compression, or, specifically, finite-context models, by obtaining a measure of similarity between a reference and a target. The method uses an approach based on finite-context models for the creation of a statistical model of the reference sequence and obtaining the estimated number of bits necessary for the encoding of the target sequence, using the reference model. NET-ASAR is freely available, under license GPLv3, at https://github.com/manuelgaspar/NET-ASAR.

Keywords: DNA sequence search · Data compression
Finite-context models · Similarity

1 Introduction

Deoxyribonucleic acid, commonly known as DNA, is the genetic material of every organism and contains all the necessary instructions for its function and development. A DNA sequence is a succession of letters, known as bases or nucleotides, with four different possibilities (A, C, G, T).

Different combinations of bases and their order allows for the great diversity between species and even within the same species. The sequences of nucleotides can be separated into coding and non-coding regions, where the relevant information to protein production can be found in the coding regions. The process of unveiling DNA nucleotides is known as DNA sequencing [1].

The sequencing process enables the analysis and study of genetic data. However, sequencing is not perfect, as, currently, it is only possible to sequence fragments. As such, it is similar to finding the order and the content of a few pieces of a huge puzzle, that we want to assemble and analyze. The pieces of the puzzle, in this case, tiny fragments of DNA (between 25 to 300 bases), might

© Springer Nature Switzerland AG 2019
F. Fdez-Riverola et al. (Eds.): PACBB 2018, AISC 803, pp. 114–122, 2019.
https://doi.org/10.1007/978-3-319-98702-6_14

also contain several errors, and, as such, the analysis of the genomic sequences needs to be performed with tools/methods that are aware of such difficulties [2].

In recent years, there has been a great increase in the amount of sequenced DNA, which has lead to a problem: the great amount of data that needs to be stored. Data compression plays a crucial part in solving this problem. By having ways of discarding redundant information, it is possible to reduce the size needed for storage. Thus, several methods for DNA compression have been proposed throughout the years (see, for example, [3–8]).

The ability to locate and identify similar sequences of DNA is extremely important in understanding, for example, the connections between different organisms and their common traits, such as common ancestors. Different tools have been implemented with this goal in mind, using different techniques and focusing on different DNA characteristics (see, for example, [9–13]).

In this paper, we intend to evaluate the use of data compression metrics, as a method to locate DNA subsequences with some degree of similarity. As such, an alignment-free [14] technique, based on three-state Finite-Context Models (FCMs) [15], was implemented.

The paper is organized as follows. In Sect. 2, we describe the method. In Sect. 3, we present some results obtained with the developed tool. In Sect. 4, we draw some final conclusions.

2 Method

The method presented in this paper is based on three-state Finite-Context Models. The main premise is that, by creating a statistical model of the reference sequence, and using said model to compute the information content of each base (in bits) belonging to the target sequence, it should be possible to observe higher compression rates when the reference sequence, or a sequence with some degree of similarity, is present in the target.

The method consists of two main steps:

1. Modeling – A set of FCMs is loaded with the information of the reference sequence that, when concluded, are kept fixed until the end.
2. Information content calculation – Using the previously loaded models of the reference, the number of bits needed to encode each base of the target sequence is calculated.

In the following subsections, we describe the method used and how it was implemented in a computational tool.

2.1 Description

DNA coding regions have an intrinsic property called three-base periodicity [16]. In order to explore this property, the proposed method is implemented, as mentioned previously, with a three-state FCM.

Consider a source of information, that outputs symbols belonging to a finite alphabet, $\mathcal{A}$, and a sequence $x^t = x_1 x_2 x_3 \ldots x_t$, that refers to the generated symbols at time t, as shown in Fig. 1.

At any given time (or symbol), the three-state FCM assigns a probability estimate to the symbols of the alphabet $\mathcal{A}$. These estimates are calculated according to a conditioning context of finite size, M, of past symbols (order-M FCM) [17,18], $c^t = x_{t-M+1}, \ldots, x_{t-1}, x_t$. The total number of possible contexts of an order-M finite-context model is, therefore, $|\mathcal{A}|^M$, which, in the scope of this work, is 4^M, since $|\mathcal{A}| = 4$ (referring to the four DNA bases). The three states of the model are selected periodically and each state can be seen as a single FCM.

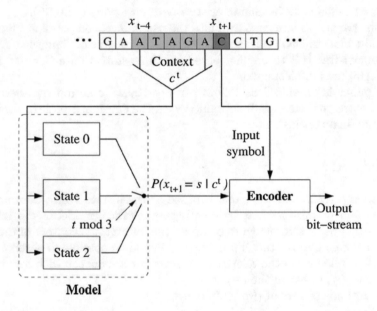

Fig. 1. Three-state FCM. The probability of the next outcome x_{t+1} is conditioned by the context depth, M, and $(t \bmod 3)$. The Encoder calculates the amount of information.

The probability of x_{t+1} not only depends on M, but also on $(t \bmod 3)$, which is used for state selection. The probability estimator is then given by

$$P(x_{t+1} = s \,|\, c^t) = \frac{{}_i n_s^t + \alpha}{\sum\limits_{a \in \mathcal{A}} {}_i n_a^t + 4\alpha}, i = t \bmod 3, \qquad (1)$$

where, in a given state, i, ${}_i n_s^t$ represents the number of past times that the symbol, s, has been generated after the conditioning context, c^t, and $\sum\limits_{a \in \mathcal{A}} {}_i n_a^t$ is the total number of times that the same context has appeared in the past.

The parameter α is a small positive value that prevents the estimation of a zero probability for symbols that have never occurred in the past.

Three-state FCMs are now easily translatable to three distinct tables that store the contexts, the number of times said contexts appear and the number of times a symbol appears given each context.

Table 1 presents a simple example of how an order-3 three-state model is usually implemented. Each row presents the last three encoded symbols (context), the counter for each symbol and the sum of all the counters (total number of times the context has appeared in the sequence). The counters are updated as each new symbol of the sequence is encoded.

Table 1. Order-3 three-state FCM example.

State 0						State 1						State 2					
c^t	n_A^t	n_C^t	n_G^t	n_T^t	$\sum n_a^t$	c^t	n_A^t	n_C^t	n_G^t	n_T^t	$\sum n_a^t$	c^t	n_A^t	n_C^t	n_G^t	n_T^t	$\sum n_a^t$
AAA	4	12	2	23	41	AAA	2	4	14	1	21	AAA	11	7	9	18	45
⋮	⋮	⋮	⋮	⋮	⋮	⋮	⋮	⋮	⋮	⋮	⋮	⋮	⋮	⋮	⋮	⋮	⋮
AGA	1	22	31	14	68	AGA	14	5	22	6	47	AGA	7	32	16	10	65
⋮	⋮	⋮	⋮	⋮	⋮	⋮	⋮	⋮	⋮	⋮	⋮	⋮	⋮	⋮	⋮	⋮	⋮
TTT	8	3	6	15	32	TTT	19	8	24	5	56	TTT	16	28	6	31	81

It is also known that the information content of the symbol x_{t+1} is given by $-\log_2 P(x_{t+1}|c^t)$ [17,18]. Therefore, the number of bits needed to encode each symbol of the sequence can be obtained from the FCM. Finally, the minimum number of bits provided by the FCMs is used.

It is important to note that events with higher probability estimates originate fewer bits to be encoded than in the case of equal probabilities (where two bits are needed).

When searching for a reference sequence, it is possible for it to be present in the target but as a reversed complement. As such, this property is considered when building the models by updating a second three-state FCM. This model is updated using the reverse complement of the context and the complement of the symbol preceding the context. Figure 2 presents a diagram summarizing how the models are updated.

In order to overcome the problem of possible desynchronization between the states of the FCMs and the target sequences, the calculations, for each base in the target sequence, are performed three times. Given the usage of two distinct three-state FCMs, the final value outputted by the method refers to the minimum value obtained by the states of both three-state models.

2.2   Implementation

NET-ASAR was implemented as command line tools, using the C++ language. NET-ASAR includes tools to map and visualize the results.

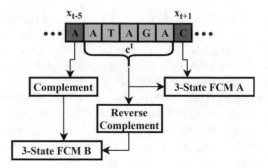

Fig. 2. Diagram showing how the two three-state models are updated.

For compiling, it was used the g++ compiler, present in Ubuntu (version 16.04) and the C++14 Standard. The method has been made publicly available, under GPLv3 license, at https://github.com/manuelgaspar/NET-ASAR.

3   Results

In this section, we present some results obtained with NET-ASAR, using synthetic and real data.

3.1   Synthetic Data

We have created a synthetic DNA sequence, consisting of (uniform) randomly generated subsequences and snippets extracted from a gene[1], with varying degrees/types of mutations. Figure 3 represents the created sequence.

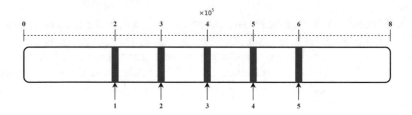

Fig. 3. Representation of the created synthetic sequence. The gene snippets are represented in red.

We have performed a search using the synthetic sequence as the target sequence and the complete gene as the reference sequence. For comparison purposes, the search was carried out with NET-ASAR and BLAST 2 Sequences [19], which is a tool that uses the BLAST [20] algorithm.

---

[1] Auxin transport protein (*BIG*).

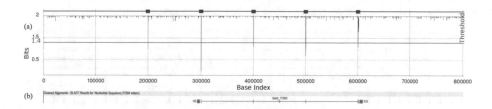

Fig. 4. Identification of the reference gene *BIG* in the synthetic target sequence using NET-ASAR (a) and BLAST (b).

Figure 4 presents the results obtained with NET-ASAR (Fig. 4a) and BLAST (Fig. 4b). We are able to observe that NET-ASAR is able to identify all the gene snippets present in the synthetic sequence, while the BLAST method fails to identify the first snippet.

3.2   Application to Ancient DNA

Using CHESTER [21,22], several potential modern-human evolutionary DNA regions, relative to two ancient genomes, were found. These regions were obtained by comparing the modern-human genome to a *Neanderthal* [23] and a *Denisovan* high coverage genomes [24], simultaneously. The analysis only mapped exact regions and, therefore, the majority of the identified regions should be present by approximate similarity, given its evolutionary role. Also, notice that two genomes do not reflect a trait of a population, but rather an indication. Besides, the ancient DNA have specific characteristics, namely the absence of several regions, mostly given by post-mortem degradation, which can be confused with evolutionary regions. For a deeper analysis, we need more archaic-human genomes, which, currently, are scarce.

After extracting the identified (potential) modern-human evolutionary regions, NET-ASAR was used to detect and locate them in three modern target organisms which share a common ancestral with the modern-human: *Pan troglodytes*, *Gorilla gorilla* and *Pongo abelii*. Since the number of identified regions was high, only some of the regions were used and, of those, only a small number will be presented in this work. The regions were selected based on two conditions: the presence of an identified and annotated gene (by NCBI) and a minimum length of 400 bases.

The identified regions were also searched in the Nucleotide collection database, using BLASTn, which is optimized for somewhat similar sequences, for comparison purposes. Table 2 presents some of the regions and the results obtained with NET-ASAR, for the three target organisms. It is important to note that some of the presented regions were not located in the Nucleotide collection database at the sames regions, mostly given chromosomal rearrangements.

Table 2. Some examples of the identified modern-human evolutionary regions and the results obtained with NET-ASAR in the target organisms. The results marked with an asterisk (*) refer to located regions that included identified genes.

Reference		Target		
Location	Gene	*Pan troglodytes*	*Gorilla gorilla*	*Pongo abelii*
C4: 121361770-121362434	QRFPR	✓*	✓*	✗
C4: 121363572-121364296		✓*	✓*	✓*
C9: 29093150-29093748	LINGO2	✓*	✓	✓
C9: 29096628-29097168		✓*	✓	✓
C9: 29095032-29095568		✓*	✓	✓
CX: 56151436-56152988	KLF8	✓*	✓	✓
CX: 56142228-56143754		✓*	✓	✓
CX: 56149828-56150760		✓*	✓	✓

Although only some examples were presented, it can be observed that the potential modern-human evolutionary regions can be located, with some degree of similarity, in three close species. The results obtained indicate that, potentially, the genes contained in the regions have high rates of evolution. Additionally, the potential of the method is apparent, since it was able to identify most of the regions, unlike BLASTn. Plots of search results obtained with NET-ASAR can be found in https://github.com/manuelgaspar/NET-ASAR.

4   Conclusions

By taking advantage of some of the characteristics of DNA, we have implemented and tested NET-ASAR, under different conditions, providing positive outcomes. One of the facts to take into account is that, while the method can provide correct results, it also demands a high degree of familiarity with the method's usage. Since the process is hard to automate, the user is required to determine the correct parameters, such as context size and filtering, needed for each search. The choice must be made having in mind the characteristics of both the reference and target sequences. It should also be noted that, even though the results of some given search can be satisfactory, the user should also analyze the identified regions closely, as false positives are a possibility.

In the future, it would be beneficial to improve the algorithm, as it is demanding in terms of computational resources. Furthermore, it could be improved by selecting parameters, such as context size, automatically, in order to easily adapt to different conditions. Additional results can be found in [25].

Acknowledgments. This work was partially funded by National Funds through the FCT - Foundation for Science and Technology (UID/CEC/00127/2013, PTDC/EEI-SII/6608/2014).

References

1. Sanger, F., et al.: DNA sequencing with chain-terminating inhibitors. PNAS **74**(12), 5463–5467 (1977)
2. Tomkins, J.: How genomes are sequenced and why it matters: implications for studies in comparative genomics of humans and chimpanzees. Answers Res. J. **4**, 81–88 (2011)
3. Chen, X., et al.: DNACompress: fast and effective DNA sequence compression. Bioinformatics **18**(12), 1696–1698 (2002)
4. Pinho, A.J., et al.: On the representability of complete genomes by multiple competing finite-context (Markov) models. PLoS ONE **6**(6), e21588 (2011)
5. Cao, M.D., et al.: A simple statistical algorithm for biological sequence compression. In: Data Compression Conference, DCC 2007, pp. 43–52. IEEE (2007)
6. Christley, S., et al.: Human genomes as email attachments. Bioinformatics **25**(2), 274–275 (2008)
7. Wang, C., Zhang, D.: A novel compression tool for efficient storage of genome resequencing data. Nucleic Acids Res. **39**(7), e45 (2011)
8. Deorowicz, S., et al.: GDC 2: Compression of large collections of genomes. Sci. Rep. **5**, 11565 (2015)
9. Altschul, S.F., et al.: Gapped BLAST and PSI-BLAST: a new generation of protein database search programs. Nucleic Acids Res. **25**(17), 3389–3402 (1997)
10. Thompson, J.D., et al.: CLUSTAL W: improving the sensitivity of progressive multiple sequence alignment through sequence weighting, position-specific gap penalties and weight matrix choice. Nucleic Acids Res. **22**(22), 4673–4680 (1994)
11. Edgar, R.C.: MUSCLE: multiple sequence alignment with high accuracy and high throughput. Nucleic Acids Res. **32**(5), 1792–1797 (2004)
12. Darling, A.E., et al.: progressiveMauve: multiple genome alignment with gene gain, loss and rearrangement. PLoS ONE **5**(6), e11147 (2010)
13. Vialle, R.A., et al.: RAFTS3: Rapid Alignment-Free Tool for Sequence Similarity Search. bioRxiv 055269 (2016)
14. Zielezinski, A., et al.: Alignment-free sequence comparison: benefits, applications, and tools. Genome Biol. **18**(1), 186 (2017)
15. Pinho, A.J., et al.: A three-state model for DNA protein-coding regions. IEEE Trans. Biomed. Eng. **53**(11), 2148–2155 (2006)
16. Trifonov, E.N., Sussman, J.L.: The pitch of chromatin DNA is reflected in its nucleotide sequence. PNAS **77**(7), 3816–3820 (1980)
17. Salomon, D.: Data Compression - The Complete Reference, 3rd edn. Springer Science & Business Media, London (2004)
18. Bell, T.C., et al.: Text Compression. Prentice Hall, Englewood Cliffs (1990)
19. Tatusova, T.A., Madden, T.L.: BLAST 2 Sequences, a new tool for comparing protein and nucleotide sequences. FEMS Microbiol. Lett. **174**(2), 247–250 (1999)
20. Altschul, S.F., et al.: Basic local alignment search tool. J. Mol. Biol. **215**(3), 403–410 (1990)
21. Pratas, D., et al.: Visualization of distinct DNA regions of the modern human relatively to a neanderthal genome. In: Iberian Conference on Pattern Recognition and Image Analysis, pp. 235–242. Springer (2017)
22. Pratas, D., et al.: Detection and visualisation of regions of human DNA not present in other primates. In: Proceedings of the 21st Portuguese Conference on Pattern Recognition, RecPad (2015)

23. Prüfer, K., et al.: The complete genome sequence of a Neanderthal from the Altai Mountains. Nature **505**(7481), 43–49 (2014)
24. Meyer, M., et al.: A high-coverage genome sequence from an archaic Denisovan individual. Science **338**(6104), 222–226 (2012)
25. Gaspar, M.: Automatic system for approximate and noncontiguous DNA sequences search. Master's thesis, Universidade de Aveiro (2017)

Can an Integrative SNP Approach Substitute Standard Identification in Comprehensive Case/Control Analyses?

Anna Papiez[1(✉)], Marcin Skrzypski[2], Amelia Szymanowska-Narloch[2],
Ewa Jassem[2], Agnieszka Maciejewska[2], Ryszard Pawlowski[2],
Rafal Dziadziuszko[2], Jacek Jassem[2], Witold Rzyman[2], and Joanna Polanska[1]

[1] Institute of Automatic Control, Silesian University of Technology,
ul. Akademicka 16, 44-100 Gliwice, Poland
anna.papiez@polsl.pl
[2] Medical University of Gdansk, ul. Debinki 7, 80-211 Gdansk, Poland

Abstract. This article describes the use of a comprehensive approach for the identification of differentiating gene related blocks of SNPs based on Fisher's p-value integration with a pooled correlation approximation. This pre-selection step is proposed as an alternative to advanced haplotype analyses for computational complexity reduction. The method, previously used for pathway regulation inference in eQTL data, is with the necessary modification especially suited for high-dimensional population genetics studies with a case/control design, where extensive numbers of SNPs are identified, leading to numerous haplotype blocks to be tested. This approach extends standard allele frequency analysis to more advanced haplotype identification. The novel method succeeds at reducing the runtime while maintaining a high level of biological result accuracy when compared against the exact test for haplotypes.

Keywords: Integrative · Correlation · Linkage disequilibrium
Haplotype · SNP · Case/control · GWAS

1 Background

The rapid development of high-throughput technologies in fields of molecular biology, such as genetics, prompts the need for constant improvements and adjustment to the analysis procedures, especially in the case of ever growing amounts of data. In this work, an alternate approach for selection of genes for haplotyping is proposed. Standard single nucleotide polymorphism (SNP) case/control data analysis pipelines involve using the exact test for population differentiation. In the case of software such as EPACTS (https://github.com/statgen/EPACTS), the standard analysis pipeline allows for indicating significant SNPs by investigation of allelic frequencies. By contrast, the advanced algorithm allows for the identification of haplotype blocks that are potential

© Springer Nature Switzerland AG 2019
F. Fdez-Riverola et al. (Eds.): PACBB 2018, AISC 803, pp. 123–130, 2019.
https://doi.org/10.1007/978-3-319-98702-6_15

risk/protection factors in a given disease or condition. The method itself is computationally complex, as it requires an iterative procedure based on Markov chains [7]. This is done by exploring the space of all possible contingency tables to determine their probability of occurrence. The proportion of these probabilities is considered an unbiased estimate of the p-value. Additionally, in many studies, such as the one discussed herewith, the gametic phase is unknown and needs to be estimated. In commonly used statistical software for population genetics analysis - Arlequin [2] - for genotypic data with unknown gametic phase, the contingency tables for the exact test are built from sample genotype frequencies [4].

The alternative novel procedure presented here is based on Fisher's p-value combination algorithm [3], which integrates differentiation strength of individual SNPs within a gene block. Satterthwaite's approximation [8] incorporating correlation resulting from linkage disequilibrium testing addresses the potential associations between SNPs present in close proximity. Additionally, introducing the pooled correlation modification makes the procedure suitable for the case/control study design. Most importantly, the advantages of this approach are based on calculation simplicity, while maintaining the significant biological results. The idea is to use the procedure as a pre-selection tool in population genetics (GWAS) case/control studies.

Differences in construction between the Markov chain simulation method and Fisher's p-value combination are depicted in Fig. 1.

2   Materials and Methods

2.1   Data

The data in this study comprised a set of SNP genotyping data obtained from exome sequencing of samples from 138 smoking individuals divided into two groups: 71 healthy controls and 67 cases of non-small cell lung cancer. The study was conducted in order to identify single nucleotide variants (SNV) linked with increased risk of lung cancer. Paired-end sequencing of 100bp fragments was carried out in Illumina HiSeq 2000 platform. Next, 180,335 genomic variants were called from aligned reads with the Genome Analysis Toolkit (GATK) [6]. The identified SNVs were checked for Minor Allele Frequency and missing values. In both cases 5% values were used as the threshold for a SNP to qualify for the missing data imputation procedure. This was conducted using fastPHASE software [9]. After these preprocessing steps a total of 81,639 SNPs was considered for further analyses. The SNPs were tested by means of the logistic score, whether they belonged to one of the three models describing association to the case/control distribution: additive, dominant or recessive. This stage completes the standard analysis pipeline. In the second advanced stage, the variants without significant deviations from Hardy-Weinberg Equilibrium in the healthy control samples and with a p-value of < 0.05 in at least one of the model types, were then assigned to haplotype blocks, according to the gene they belonged to, i.e., a variant had to be a single nucleotide polymorphism and part of the

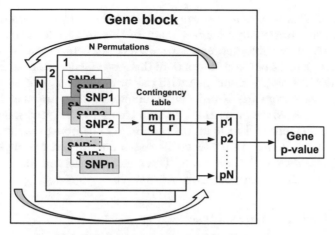

(a) Exact test for sample differentiation

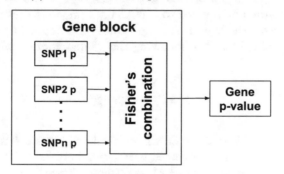

(b) Fisher's combined p-value algorithm

Fig. 1. Diagrams representing the principles behind the two algorithms: (a) exact test used in the advanced analysis pipeline and (b) the proposed Fisher's combined p-value novel pipeline.

exome sequence (not labelled as either of: Insertion/Deletion/Intergenic/Intron). There remained 3,290 polymorphisms fulfilling these criteria and, thus, assigned to haplotype blocks. Only genes with a minimum of 2 SNPs fulfilling the criteria were considered in the downstream analyses and there was a total of 587 such haplotype blocks.

2.2 Advanced Analysis Approach

The standard approach to identify haplotype blocks acting as potential biomarkers of lung cancer is to perform testing for sample haplotype frequency differentiation. For this purpose the data on haplotype blocks were submitted to the Arlequin 3.5 software package [2]. More specifically, the exact test of sample differentiation based on genotype frequencies tool was chosen to be performed on all blocks. This test relies on the Markov chain iterative algorithm, which makes it a computationally complex task. In this setting for each gene block 100,000 contingency table permutations were constructed with 10,000 burn-in steps. The frequencies of these blocks were then used again to assess if there exist any particular haplotypes associated with risk or protection. This was performed using components from the G-test corresponding to the particular haplotypes. The underlying assumption is that if there exists a dominant haplotype, the associated component will have an outstanding G component value associated. The rule for a component to be deemed outstanding was that it appeared as an outlier in Dixon's detection criterion. These gene blocks are finally subject to haplotype assignment using the ELB method [2].

2.3   Novel Integrative Analysis

The integrative approach proposed here is based on Fisher's p-value combination method. This algorithm, applied to GWAS data previously with a different purpose [1], transforms the logistic score p-values (p_i) assigned to L SNPs in order to obtain a combined statistic following the χ^2 distribution with $2L$ degrees of freedom:

$$Z = \sum_{i=0}^{L} -2log(p_i) \tag{1}$$

However, this holds in the case where p-values are associated with independent features. In these circumstances, correlation between SNPs occurring in the same genes has to be dealt with. For this purpose, Satterthwaite's approximation accounting for this correlation is used:

$$Z = \sum_{i=0}^{L} -2log(p_i) \sim \alpha\chi_g^2 \tag{2}$$

In this case α and g are obtained with the following formulas, where ρ_{ij} is a measure of correlation between two SNPs:

$$\alpha = 1 + \frac{2\sum_{i<j} \rho_{ij}}{L} \tag{3}$$

$$g = \frac{2L}{\alpha} \tag{4}$$

As described in [1], the correlation is estimated based on the R^2 value of linkage disequilibrium determination. Nevertheless, the experimental design here required another modification, i.e. pooling the correlation coefficients from cases and controls. For this aim, Hedges and Olkin pooled correlation coefficients based on homogeneity testing were calculated [5].

$$\rho_{ij} = pooled(\rho_{ij}^{controls}, \rho_{ij}^{cases}) \tag{5}$$

The final combined p-values were assigned a dynamic significance threshold, that is, the threshold was also a combined p-value corresponding to single p-values at 0.1 (0.05 for one-sided test). Due to the dynamic nature of the threshold, direct comparison between the combined p-values cannot be done and multiple testing correction cannot be applied.

The flow of the standard, advanced and novel integrative analysis pipelines is illustrated in Fig. 2.

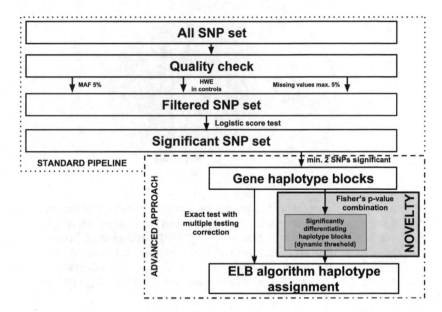

Fig. 2. Block diagram showing the gradually increasing advancement of the pipeline. The top presents a standard single SNP allele frequency analysis. Further down, the advanced haplotype block identification is introduced leading to a comprehensive results from SNP interactions within blocks. The novel integrative method enables reduction of haplotype assignment procedures, by identifying blocks of SNPs using the combined p-value as a threshold.

3   Results

3.1   Quantitative Results

Firstly, by means of the advanced approach exact test, 176 haplotype blocks were indicated as differentiating cases and healthy controls, and with Benjamini-Hochberg correction for multiple testing, 16 blocks in total were retained. The final step of checking for outstanding dominant haplotypes, based on Dixon's outlier criterion, yielded 8 dominant haplotypes, which became the input for Arlequin's haplotype frequency estimation tool.

In the course of integrative p-value combination pre-selection, 332 out of the 587 total genes fell into the significant category in terms of the dynamically assigned thresholds. This set contained 136 of the 176 differentiating haplotypes identified in the standard approach. Within this group, seven of the eight genes with outstanding dominant haplotypes were present. The intersection of genes being the outcome of the two approaches is presented in Fig. 3.

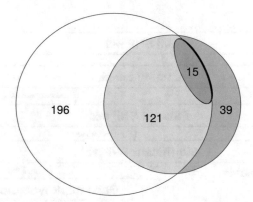

Fig. 3. Venn diagram presenting the comparison between differentiating gene blocks identified using the advanced approach (in grey) and novel integrative approach (in white). The inner ellipse presents the common 15 genes identified using both approaches out of 16 genes remaining after multiple testing correction in the advanced approach. The thin black area indicates the one gene not identified by the integrative approach. Diagram produced by means of the *eulerr* package tool (https://github.com/jolars/eulerr).

3.2   Runtimes

Calculations for both approaches were carried out in a sequential processing scheme on an Intel Core i5-3320M processor, 2.60 GHz with 16 GB DDR3 1600 MHz RAM memory. The computations were run in sequential mode in order to enable direct comparison between total runtime for the total 587 genes. The total runtimes are presented in Table 1.

Table 1. Presentation of the differences in runtimes for the advanced and novel integrative pipelines.

	Total runtime [s]	Average runtime per gene [ms]
Exact test for differentiation	185.65	316.27
Fisher's p-value combination	3.11	5.30

4   Discussion

The idea behind this work was to present an alternative technique for haplotype block pre-selection in genetics studies, which would provide a simple approach for primary reduction of features to be processed in later stages (especially time consuming such as haplotype assignment). The integrative method based on Fisher's p-value combination seems to retain the conditions fulfilled while using the exact test for differentiation, in that:

- it does not require phasing of the haplotypes,
- differentiation of a haplotype is based on differentiation of individual SNPs in the entire block.

Satterthwaite's modification relying on the incorporation of correlation between the tested features, takes into account the possible interactions between SNPs in a block, even more so, as the correlation metric is approximated by R^2 from linkage disequilibrium testing. The main advantage of this integrative approach is the straightforwardness of the computations required, as opposed to the Markov chain simulations necessary in case of performing the exact test. Indeed this gain may be observed at the level of runtimes, as the integrative testing is completed in two orders of magnitude less time per block than the standard approach (Table 1).

When investigating the quantitative results, 332 genes are identified as potentially significantly differentiating in the integrative approach with a dynamic p value threshold, as opposed to 176 genes in the standard approach. In this way the integrative method may be regarded as less conservative, which in some studies may be of particular advantage, as the p-value threshold is not set arbitrarily, but rather adaptively according to the test's power. Therefore, this may prevent the loss of information due to the necessity of deciding on a fixed threshold for the exact test. In comparison between the two sets of genes, the integrative approach set contains a strong majority of genes indicated in the standard approach (Fig. 3). Moreover, nearly all of the genes, but for one, are present in the set identified by standard approach after multiple testing. This shows that the integrative pre-selection pipeline allows for computational complexity reduction, while preserving a high level of biological results found with the use of a standard pipeline.

5   Conclusions

In this work, an algorithm for pre-selection of differentiating haplotype gene blocks is proposed as an alternative to the exact test. This is a method that serves as an extension of the standard allele frequency SNP differentiation analysis performed by means of commonly used software such as EPACTS. In this comprehensive approach, the appropriate models are identified for SNPs and haplotype blocks alike, which enables investigation into SNP interactions within a block. In the examined study, the novel integrative approach successfully indicates a majority of the biological results identified by the advanced approach with low computational complexity. The pooled variance modification enables the application to case/control experimental designs. This step in the analysis pipeline may be particularly useful before conducting analyses on numerous SNPs and gene blocks, especially in the case of arduous downstream tasks, such as ELB haplotype assignment. For the sake of the method's development, further studies may be conducted towards exploring p-value combination methods applied in terms of algorithm improvement. Moreover, it would be worth to examine how combining SNPs and haplotype blocks into the discriminating models could affect the final results by introducing the possibility of such mixed interactions.

Acknowledgements. This work was funded by The Polish National Centre for Research and Development grant no. PBS3/A7/29/2015/ID-247184 (AP) and National Science Centre, Poland grant no. 2015/19/B/ST6/01736 (JP). Calculations were carried out using GeCONiI infrastructure (POIG02.03.01-24-099).

References

1. Cui, Y., Li, S., Williams, B.L.: A combined p-value approach to infer pathway regulations in eQTL mapping. Stat. Interface **4**(3), 389–401 (2011)
2. Excoffier, L., Laval, G., Schneider, S.: Arlequin (version 3.0): an integrated software package for population genetics data analysis. Evol. Bioinform. **1**, 47–50 (2005)
3. Fisher, R.A.: Statistical methods for research workers. In: Kotz, S., Johnson, N.L. (eds.) Breakthroughs in Statistics, pp. 66–70. Springer, New York (1992)
4. Goudet, J., Raymond, M., de Meeüs, T., Rousset, F.: Testing differentiation in diploid populations. Genetics **144**(4), 1933–1940 (1996)
5. Hedges, L., Olkin, I.: Statistical Methods for Meta-Analysis. Elsevier Science, New York (2014)
6. McKenna, A., Hanna, M., Banks, E., Sivachenko, A., Cibulskis, K., Kernytsky, A., Garimella, K., Altshuler, D., Gabriel, S., Daly, M., et al.: The Genome Analysis Toolkit: a MapReduce framework for analyzing next-generation DNA sequencing data. Genome Res. **20**(9), 1297–1303 (2010)
7. Raymond, M., Rousset, F.: An exact test for population differentiation. Evolution **49**(6), 1280–1283 (1995)
8. Satterthwaite, F.E.: An approximate distribution of estimates of variance components. Biom. Bull. **2**(6), 110–114 (1946)
9. Scheet, P., Stephens, M.: A fast and flexible statistical model for large-scale population genotype data: applications to inferring missing genotypes and haplotypic phase. Am. J. Hum. Genet. **78**(4), 629–644 (2006)

Simple Pattern-only Heuristics Lead to Fast Subgraph Matching Strategies on Very Large Networks

Antonino Aparo[1], Vincenzo Bonnici[1], Giovanni Micale[2], Alfredo Ferro[2], Dennis Shasha[3], Alfredo Pulvirenti[2], and Rosalba Giugno[1(✉)]

[1] University of Verona, Verona, Italy
rosalba.giugno@univr.it
[2] University of Catania, Catania, Italy
[3] New York University, New York, USA

Abstract. A wide range of biomedical applications entails solving the subgraph isomorphism problem, i.e. finding all the possible subgraphs of a target graph that are structurally equivalent to an input pattern graph. Targets may be very large and complex structures compared to patterns. Methods that address this NP-complete problem use heuristics. Their performance in both time and quality depends on a few subtleties of those heuristics. This paper compares the performance of state-of-the-art algorithms for subgraph isomorphism on small, medium and very large graphs. Results show that heuristics based on pattern graphs alone prove to be the most efficient, an unexpected result.

Keywords: Subgraph isomorphism · Networks biology
Search strategy

1 Introduction

In the last decade, significant national, international and private research resources have been directed at data-driven genome-level projects, such as the 1000 Genomes Project[1], Encyclopedia of DNA Elements Project [2], and The Cancer Genome Atlas Project (TGCA)[3]. Different approaches have been proposed to integrate data across multiple omics (i.e. whole species) data sets [11]. Well-known examples are protein-protein interaction networks, genetic regulatory networks, and metabolic networks. Analysis of such networks has lead to the discovery of recurrent and statistically over-represented sub-networks [8,10].

Such biomedical applications entail solving the subgraph isomorphism problem, an NP-complete problem [9]. The problem consists of finding all the possible subgraphs of a reference graph (called target) that are structurally equivalent to

---

[1] http://www.internationalgenome.org/.
[2] http://www.encodeproject.org.
[3] https://cancergenome.nih.gov/.

© Springer Nature Switzerland AG 2019
F. Fdez-Riverola et al. (Eds.): PACBB 2018, AISC 803, pp. 131–138, 2019.
https://doi.org/10.1007/978-3-319-98702-6_16

another graph (called the pattern)[2]. Given the potentially exponential problem complexity and the ever growing target graphs, heuristics for subgraph isomorphism algorithms must achieve scalability for large networks [1,5].

We have focused on the two most scalable graph searching algorithms, RI [2,3] and VF3 [4]. In the literature, such algorithms have been compared on graphs up to 10000 vertices [4]. In the present paper, we have generated a benchmark of graphs containing 1008 target graphs of size up to 20.000 vertices using three different graph generation models (Erdös-Rényi, Barabási, Forest Fire), and about 150.000 pattern graphs of different sizes and densities. Our results show that heuristic strategies based only on patterns perform well on small graphs and perform considerable better on large graphs w.r.t strategies which account also the target graphs.

2   An Annotated Overview of Subgraph Isomorphism Algorithms

Basic notions and problem definition. A graph G is a pair (V, E), where V is the set of vertices and $E \subset V \times V$ is the set of edges connecting these vertices. G is *labeled* when a set of labels A is assigned to the vertices and/or the edges of G using an injective function: a vertex label function $\alpha : V \to A$ and/or an edge label function $\beta : E \to A$. Labeled graphs can be represented by a quadruple (V, E, α, β). A graph is *directed* if each edge has a direction from the source vertex to the destination vertex, otherwise the graph is *undirected*. *Dense graphs* are those for which the ratio $|E| \, / \, |V|$ is relatively high (e.g. the number of edges approaches the square of the number of vertices), where $|V|$ and $|E|$ are the cardinalities of the two sets. Otherwise, graphs are considered *sparse*. Given a *pattern graph* $G_p = (V_p, E_p, \alpha_p, \beta_p)$ and a *target graph* $G_t = (V_t, E_t, \alpha_t, \beta_t)$ the subgraph isomorphism problem is to find a injective function $f : V_p \to V_t$, mapping each vertex of G_p to a unique vertex of G_t, while satisfying the following conditions: (i) $\forall v \in V_p f(v) \in V_t$ (ii) $\forall u, v \in V_p : u \neq v \Rightarrow f(u) \neq f(v)$; (iii) $\forall (u, v) \in E_p \Rightarrow \exists (f(u), f(v)) \in E_t$; (iv) $\forall v \in V_p \Rightarrow \alpha_p(v) = \alpha_t(f(v))$ *and* $\forall (u, v) \in E_p \Rightarrow \beta_p(u, v) = \beta_t(f(u), f(v))$. This is an injective mapping because the target graph may have edges that the pattern graph doesn't. As mentioned in the introduction, the subgraph isomorphism problem is NP-complete [9].

Features of subgraph isomorphism algorithms. Heuristic algorithms to solve the subgraph isomorphism problem build on two main paradigms: tree search ([3,4]) and constraint programming ([7,12,13]). In the *Tree Search* approach, for each pattern graph, a tree is formed containing all possible mappings of the pattern into the target graph. A path from the dummy root to a leaf contains $|V_p|$ nodes (i.e. the number of nodes in the pattern), each node in the tree is a mapping $(v, f(v))$ with $v \in V_p$. A path from the root to an internal node represents a partial matching, because not all vertices in the pattern graph have been mapped up to that point. The tree is obtained by incrementally extending a partial solution with a new pair of vertices $(v, f(v))$ that satisfies the subgraph isomorphism constraints. If there are no more candidate pairs, the algorithm backtracks and prunes that branch. An advantage of this approach is that the algorithms have linear memory complexity with respect to the number of vertices.

The *constraint programming* approach is based on the pre-computation of compatibility domains, i.e. pattern vertices that are associated to sets of target vertices that satisfy subgraph isomorphism constraints. These pre-computed domains are used during the matching phase to select candidate pairs. At each matching step, domains are reduced by propagating the constraints from the matched vertices to the unmatched ones and filtering away those vertices that can not be contained in a final solution. The main problem with this approach is that it consumes a lot of memory, making it not scalable on large graphs.

In summary, Single Search Tree (SST) approaches offer better performance and scalability than constraint programming based approaches [2]. The most important heuristic choices of SST-based algorithms are the variable ordering and how the tree is visited.

Variable ordering, also known as the search strategy, is the determination of an ordering of the pattern graph vertices in the branches of the search tree. The order can be *static* or *dynamic*. Static ordering means that the ordering is fixed a priori, before the search phase starts, and remains the same for all paths through the tree. In dynamic ordering, the ordering can change for different branches, implying additional computation during the traversal of the tree. Search strategies differ depending on which properties and which graphs are examined. An algorithm can consider features only of the target graph (*target-dependent*) [6] or only of the pattern graph (*pattern-dependent*) [3] or both [4]. State-driven variable ordering [3,4] is based on the current state of computation, while domain-driven [6] regards the values in the variable domains.

The best-performing (and most recent) algorithms using an SST approach are VF3 [4] and RI [2,3].

VF3 adopts a *dynamic, target-dependent* and *pattern-dependent* variable ordering. A scoring function gives a priority to the vertices that have more constraints to satisfy, taking into consideration also the probability of finding a compatible vertex in the target graph. The algorithm in the preprocessing phase partially pre-computes sets, called *feasibility sets*, for each vertex. Given a state s of the SST, during the matching process, classification functions are used to subdivide the feasibility sets, so that if two nodes belong to different classes they will never be coupled together. VF3 uses *look-ahead* procedures to predict downward inconsistency, applying constraints and properties related to the current partial matching to the neighbors of a vertex.

RI uses a *static* variable ordering and forms the tree in a *target-independent* way. Static variable ordering strategy allows RI to reduce the search space without using predictive pruning verification (which can be expensive). The vertices are sorted by a score function that takes into account the neighbors of a vertex in the pattern graph alone. The goal of the ordering is to start with the parts of the pattern graph that must satisfy the most possible constraints, so can be filtered away most quickly. During the matching phase, the vertices of the pattern graph are compared with those of the target graph using the search tree. As the search tree is traversed the subgraph isomorphism conditions are tested. If they are not satisfied, then RI *backtracks* (cutting the corresponding branch and reducing the search space).

3   Results

In order to compare RI and VF3 across a variety of scenarios, we have generated a dataset containing graphs obtained according to stochastic (*Erdös-Rényi*) and scale-free models (*Barabási* and *Forest-Fire*), starting from small (100 node) and sparse graphs up to large (20,000 node) and dense graphs both with few or many labels. We used the python library *network* (for Erdös-Rényi graphs) and *igraph* (for Bababasi and Forest-Fire graphs). Altogether, our benchmark consists of 1,008 target graphs.

After the generation of target graphs, a *random walk based algorithm* was used to extract the patterns by varying pattern size and density. Starting from a vertex of the target graph, a neighbor is selected randomly, until the desired number of vertices of the pattern is reached. Subsequently the algorithm adds to the pattern the edges not yet selected among the chosen vertices, until the specified density is reached.

Tests were running on a machine equipped with an Intel Core i7-7700 3.60 GHz 8-core CPU and 15 GB of RAM running a Linux Ubuntu 17.04 64bit operating system. A 3 min timeout has been set for both algorithms.

Comparisons on the Erdös-Rényi dataset. A graph G with parameter p is created by connecting its N vertices randomly. Each edge is included in the graph with probability p independently of any other edge. $E = pN(N-1)$ is the expected number of edges. This dataset contains 464 targets graphs of 100, 200, 500, 1,000, 2,000, 5,000, 10,0000, 20,0000 vertices with labels ranging from 0.1% to 30% depending on the number of vertices and with probability p: 0.001, 0.002, 0.005, 0.01, 0.02, 0.05, 0.1, 0.2, 0.3, 0.4. From these targets, up to 240 pattern graphs were extracted for each target with 4, 8, 16, 24, 32, 64, 128, 256 number of vertices and three different pattern densities: 0.1 (sparse), 0.5 (medium dense), and ≈ 1 (dense). Figure 1 reports running times of the two algorithms on the *Erdös-Rényi* dataset. Times are grouped by the number of target vertices of the matching instances (Fig. 1(a)). The majority of the RI times are below the VF3 boxes and few outliers have comparable execution times with VF3. The other three plots show running times on target graphs having 10k vertices. Both algorithms increase the time requirement by incrementing the number of target vertices and the value of the density parameter p. RI shows a lower dependence to these factors. For example, for the maximum value of p, VF3 triples its average running time, while the time of RI is less affected. Times for $p > 0.2$ were not

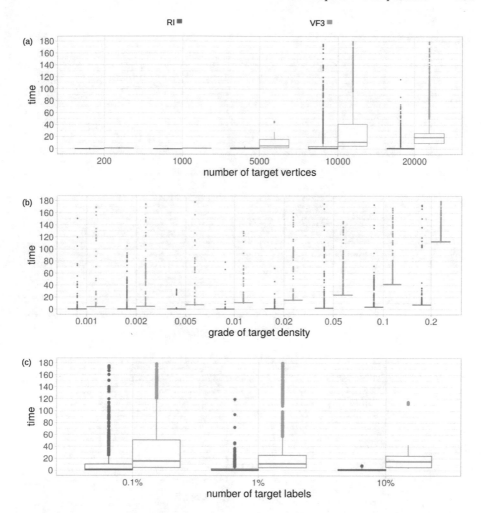

Fig. 1. Comparison of RI (red) and VF3 (green) on the Erdös-Rényi dataset. Execution time is expressed in seconds. (a) Scalability of the algorithms on the size of the target graph, the x-axis reports the number of vertices of the graph. (b) Scalability of the algorithms as a function of the density of the target graph having 10k vertices. Here, the x-axis the probability p of the Erdös-Rényi models. (c) Performance of the two algorithms on the number of labels in the target graphs having 10k vertices, the labels express the percentage with respect to the number of vertices.

reported since for graphs having 10k or 20k nodes, VF3 went in timeout the 100% of cases whereas RI only the 35%. The dependence of both algorithms on the percentage of target labels is shown in Fig. 1(c). The variation in execution time of RI is less than that of VF3, and average times are faster using RI. In general, as expected, the greater the number of distinct labels the less the time required, because fewer target vertices are compatible with the pattern vertices.

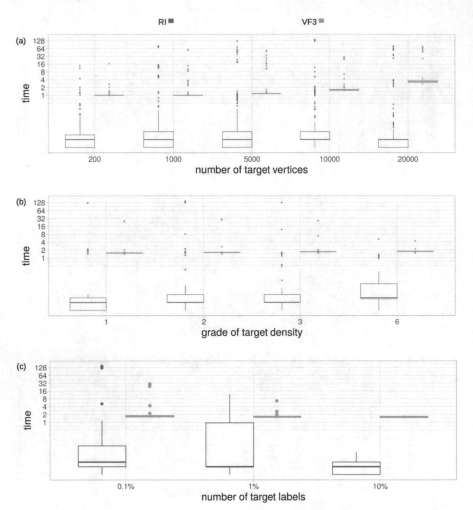

Fig. 2. Comparison of RI (red) and VF3 (green) on the *Barabási-Albetr* dataset. Execution time is expressed in seconds. (a) Scalability of the algorithms on the size of the target graph, the x-axis refers to the number of vertices of the graph. Times are expressed in log scale. (b) Scalability of the algorithms on the density of the target graph having 10k vertices, the x-axis the parameter m of the Barabási-Albert models. (c) Performance of compared algorithms on the number of labels in the target graphs having 10k vertices, the labels express the percentage w.r.t. the number of vertices.

Comparisons on the Barabási dataset. Genetic networks may show complex topology, containing few *hubs*, i.e. vertices with a high number of interactions compared to the rest of the network vertices. For example, in a metabolic network, hubs are molecules like ATP or ADP, energy carriers involved in a large number of chemical reactions. A model based on these characteristics reproduces the observed stationary *scale-free distributions* similar to the *Barabási-Albert*

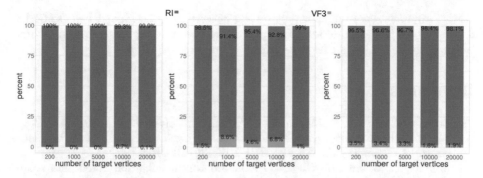

Fig. 3. Number of times, expressed as percentages w.r.t. the total amount of tests, that one algorithm (RI is red and VF3 is green) has been faster than the other. The results are grouped by number of target vertices for the Erdös-Rényi, Barabási-Albert and Forest Fire models, respectively from the left to the right.

one. The network starts with m_0 vertices. At each time step a new vertex is added to the network whit $m (\leq m_0)$ links that connect it to existing vertices in the network. The probability p_i that a link of the new vertex connects to an existing vertex i is proportional to the degree k_i of i $p_i = \frac{k_i}{\sum_j k_j}$; where the sum is made over all existing vertices j. After t time steps, the Barabási-Albert model generates a network with $N = t + m_0$ vertices and $m_0 + m_t$ links.

This dataset has 384 target graphs having 200, 500, 1,000, 5,000, 10,000, 20,000 vertices, different numbers of outgoing edges (1, 2, 3, 6) and different numbers of labels in relation to the number of vertices (0.1%, 1%, 10%). 28,800 patterns were extracted by varying the number of vertices (4, 8, 24, 32, 64) and density (0.1, 0.5, ≈1). Results are shown in Fig. 2. Times are grouped as in the Fig. 1. Running times are faster than in the previous dataset, thus in Fig. 2(a) we report times in log scale. RI shows a general behavior faster than the competitor and less dependence from the target features variability.

Due to space constraints, we don't report and picture details on the Forest-Fire model datasets. Briefly, the obtained results for the Forest-Fire model are similar to those for the Barabási-Alber model, but, both algorithms have a few more outliers. Figure 3 shows that RI generally outperforms VF3 on small, medium and very large graphs. Charts show the number of times the tested algorithms are faster than the competitor. They are built by taking into account every combination of target and pattern graph on the three models, and the amounts are expressed as a percentage of wins on the total tests. VF3 reaches its maximum of wins the 8.6% instances on the Barabási-Alber model with 1k target vertices.

4 Conclusion

Fast algorithms for subgraph isomorphism depend on a variety of heuristics, some of which have become quite sophisticated. This paper has shown that a simple approach that depends only on the pattern graph such as RI works better especially for large, dense target graphs where the running time of the matching becomes larger. This is of course a field of ongoing research.

References

1. Bonnici, V., Busato, F., Micale, G., Bombieri, N., Pulvirenti, A., Giugno, R.: Appagato: An approximate parallel and stochastic graph querying tool for biological networks. Bioinformatics **32**(14), 2159–2166 (2016). Cited by: 2
2. Bonnici, V., Giugno, R.: On the variable ordering in subgraph isomorphism algorithms. IEEE/ACM Trans. Comput. Biol. Bioinf. **14**(1), 193–203 (2017)
3. Bonnici, V., Giugno, R., Pulvirenti, A., Shasha, D., Ferro, A.: A subgraph isomorphism algorithm and its application to biochemical data. BMC Bioinf. **14**(Suppl 7), S13 (2013)
4. Carletti, V., Foggia, P., Saggese, A., Vento, M.: Challenging the time complexity of exact subgraph isomorphism for huge and dense graphs with vf3. IEEE Trans. Pattern Anal. Mach. Intell. **PP**(99), 1 (2017)
5. Giugno, R., Bonnici, V., Bombieri, N., Pulvirenti, A., Ferro, A., Shasha, D.: Grapes: a software for parallel searching on biological graphs targeting multi-core architectures. PLoS ONE **8**(10) (2013). Cited by: 10
6. Haralick, R.M., Elliott, G.L.: Increasing tree search efficiency for constraint satisfaction problems. Artif. Intell. **14**(3), 263–313 (1980)
7. McGregor, J.: Relational consistency algorithms and their application in finding subgraph and graph isomorphisms. Inf. Sci. **19**(3), 229–250 (1979)
8. Micale, G., Giugno, R., Ferro, A., Mongiovì, M., Shasha, D., Pulvirenti, A.: Fast analytical methods for finding significant labeled graph motifs. Data Min. Knowl. Discov. **32**, 504–531 (2017)
9. Michael, R.G., David, S.J.: Computers and Intractability: A Guide to the Theory of NP-Completeness, pp. 90–91. WH Free. Co., San Francisco (1979)
10. Milo, R., Shen-Orr, S., Itzkovitz, S., Kashtan, N., Chklovskii, D., Alon, U.: Network motifs: simple building blocks of complex networks. Science **298**(5594), 824–827 (2002)
11. Palsson, B., Zengler, K.: The challenges of integrating multi-omic data sets. Nat. Chem. Biol. **6**, 787–789 (2010)
12. Solnon, C.: Alldifferent-based filtering for subgraph isomorphism. Artif. Intell. **174**(12), 850–864 (2010)
13. Ullmann, J.R.: Bit-vector algorithms for binary constraint satisfaction and subgraph isomorphism. J. Exp. Algorithmics **15**, 1.6:1.1-1.6:1.64 (2011)

In Silico Predictions for Fucoxanthin Production by the Diatom Phaeodactylum Tricornutum

Claudia M. Bauer[1]([⊠]), Paulo Vilaça[2], Fernanda Ramlov[1],
Eva Regina de Oliveira[1], Débora Q. Cabral[1], Caroline Schmitz[1],
Rafaela Gordo Corrêa[3], Miguel Rocha[4], and Marcelo Maraschin[1]

[1] Plant Morphogenesis and Biochemistry Laboratory,
Federal University of Santa Catarina, Florianopolis, Brazil
clamonab@gmail.com
[2] Silico Life Lda., Braga, Portugal
[3] Laboratory of Phycology, Federal University of Santa Catarina,
Florianopolis, Brazil
[4] Centre Biological Engineering, School of Engineering,
University of Minho, Braga, Portugal
mrocha@di.uminho.pt

Abstract. Diatoms and brown seaweeds are the main producers of fucoxanthin, an oxy-carotenoid with important biological functions related to its antioxidative properties. The diatom *Phaeodactylum tricornutum* appears in this scenario as a good source for fucoxanthin production. Its whole genome sequence was published in 2008, and some genome-scale metabolic models are currently available. This work intends to make use of the two most recent genome-scale metabolic models published to predict ways to increase fucoxanthin production, using constraint-based modeling and flux balance analysis. One of the models was completed with 31 downstream reactions of the methylerythritol 4-phosphate plastidic (MEP) pathway. Simulations and optimizations were performed regarding inorganic carbon and nitrogen sources in the two models and comparisons were made between them. Biomass growth was predicted to increase in all sources tested, i.e., CO_2, HCO_3^-, NO_3^- and urea. However, the best results were obtained by combining CO_2 plus HCO_3^- regarding inorganic carbon, and for urea as a nitrogen source, in both models tested. As a result of optimizations for fucoxanthin production, many of the knockout reactions brought on are involved in the metabolism of pyruvate, glutamine/glutamate and nitrogen assimilation.

Keywords: Fucoxanthin · *P. tricornutum* · Metabolic engineering Constraint-based modeling

1 Introduction

Diatoms and brown seaweeds are the main producers of fucoxanthin, an oxy-carotenoid with relevant biological functions (e.g., antioxidative, antitumoral, neuroprotective, anti-obesity and anti-inflammatory activities) [1, 2]. The diatom *Phaeodactylum*

© Springer Nature Switzerland AG 2019
F. Fdez-Riverola et al. (Eds.): PACBB 2018, AISC 803, pp. 139–148, 2019.
https://doi.org/10.1007/978-3-319-98702-6_17

tricornutum appears in this scenario as a good source, even commercial, for fucoxanthin production [3]. This species' whole genome sequence was published in 2008 [4], providing information for exploring biological engineering tools. Diatoms have particular features in their metabolism due to its capacity of light harvesting for inorganic carbon fixation and also due to its secondary endosymbiotic origin which created chloroplasts with four membranes. However, these organisms are also able to grow in the absence of light (heterotrophic or mixotrophic conditions).

Some metabolic models are currently available for *P. tricornutum*. A model developed in 2008 [4] served as foundation for the first genome-scale model for this species, which in turn gave origin to a database named DiatomCyc [5]. Further, the two most recent genome-scale metabolic networks for *P. tricornutum* were published in 2016 [6, 7]. The model published in 2016 by Kim et al. [6] predicted intracellular fluxes under autotrophic, mixotrophic, and heterotrophic growth conditions, accounting for 849 reactions and 587 metabolites distributed among five compartments, i.e., cytosol, matrix of mitochondria, stroma and lumen of chloroplast, and peroxisome. Concerning fucoxanthin production, this model lacks downstream reactions of the MEP pathway which leads to fucoxanthin production. A second study [7] developed two genome-scale metabolic models, where one of them comprises a more complete range of lipid metabolism, being that both contain downstream reactions that lead to fucoxanthin formation and also gene-reaction (GR) associations, allowing phenotype predictions at the gene level.

As far as today, many studies involving *P. tricornutum* focus on lipid biosynthesis due to its importance for biofuels production. In this sense, N_2 starvation/depletion is the most studied condition for enhancing lipid production in diatoms which has been reported to induce accumulation of lipids and a decrease in fucoxanthin production by this species [8]. This may indicate that despite their common apolar nature and the same metabolic precursors, lipids and fucoxanthin diverge in some point of their metabolism in relation to nitrogen consumption. Thus, this work intents to make use of the two genome-scale metabolic models [6, 7] to predict ways to increase fucoxanthin production, using flux balance analysis (FBA). The results obtained by both models will be compared and discussed.

2   Methods

In a first step, the genome-scale metabolic model for *P. tricornutum* published by Kim et al. [6] was used to perform phenotype predictions and strain optimization towards fucoxanthin production by this species. This genome scale metabolic model already encompasses 19 initial reactions of the MEP pathway, which lead from pyruvate and glyceraldehyde-3-phosphate (G3P) combination to the geranylgeranyl diphosphate (GGDP) formation. Other 31 reactions were searched in the Kyoto Encyclopedia of Genes and Genomes (KEGG) database (http://www.genome.jp/kegg) and DiatomCyc (http://www.diatomcyc.org) and added to this model to complete MEP metabolic pathway towards the synthesis of fucoxanthin (Table 1).

Table 1. Equivalent reactions of the two models tested regarding the MEP pathway. Differences are highlighted in bold. Reactions of the model of Kim et al. [6] with the suffix "ADDED" were included in the model during this work.

Reaction ID		Products
Kim et al. [6]	Levering et al. [7]	
R_DXS	R_DXPS_h	1-Deoxy-D-xylulose 5-phosphate
R_DXR	R_DXR_h	2-C-Methyl-D-erythritol 4-phosphate
R_ISPD	R_CMS_h	4-(Cytidine 5' -diphospho)-2-C-methyl-D-erythritol
R_ISPE	R_CMK_h	(4-(Cytidine 5' -diphospho)-2-C-methyl-D-erythritol)/ (2-Phospho-4-(cytidine 5' -diphospho)-2-C-methyl-D-erythritol)
R_ISPF	R_MECDPS_h	2-C-Methyl-D-erythritol 2, 4-cyclodiphosphate
R_ISPG	R_HMBDPO_h	1-Hydroxy-2-methyl-2-butenyl 4-diphosphate
R_HDS_ADDED	R_HMBDPO_h	1-Hydroxy-2-methyl-2-butenyl 4-diphosphate
R_ISPH	R_IDS1_h	Isopentenyl diphosphate
R_ISPH2	R_IDS1_h	Isopentenyl diphosphate
R ISPH [1]	R_IDS2_h	Dimethylallyl diphosphate
R_ISPH2 [1]	R_IDS2_h	Dimethylallyl diphosphate
R_IDI	R_IDIH_h	(Isopentenyl diphosphate) (Dimethylallyl diphosphate)
R_FDPS	R_GPPS_c	Geranyl diphosphate
R_FDPS2	R_FPPS_h	*trans*, *trans*-Farnesyl diphosphate
R_GGDP	R_GGPS_c; R_GGPS_h	Geranylgeranyl diphosphate
R_T_GGDP_CP		**Geranylgeranyl diphosphate transport**
R_ISPDPPC_C_ADDED	R_IPDPt_m(cit <>mit);	Isopentenyl diphosphate transport
R_IDI-_P	R_IDIH_h	(Isopentenyl diphosphate) (Dimethylallyl diphosphate)
R_DMPPPC_ADDED	R_DMPPt_h	Dimethylallyl diphosphate transport
R_FDPS_P	R_GPPS_h	Geranyl diphosphate
R_FDPS2_P		**(Isopentenyl diphosphate) (*trans*, *trans*-Farnesyl diphosphate)**
	R_FPPS_h	***trans*, *trans*-Farnesyl diphosphate**
R_GGDP_P	R_GGPS_c	Geranylgeranyl diphosphate
	R_PSY_h	**all-*trans*-Phytoene**

(*continued*)

Table 1. (*continued*)

Reaction ID		Products
Kim et al. [6]	Levering et al. [7]	
R_PSY_ADDED		**Prephytoene-diphosphate**
R_PSY2_ADDED		**15-*cis*-phytoene**
R_PSY3_ADDED		**15-*cis*-phytoene**
R_PSY4_ADDED		**all-*trans*-Phytoene**
R_PDS1_ADDED		**9, 15, 9-tricis-zeta-Carotene**
R_PDS2_ADDED		**Phytofluene**
R_PDS3_ADDED		**all-*trans*-zeta-Carotene**
R_PDS4_ADDED	R_PDS1_c	all-*trans*-Phytoene
R_PDS5_ADDED	R_PDS2_h	all-*trans*-zeta-Carotene
	R_PHYPQOX_h	**9, 15, 9-tricis-zeta-Carotene**
	R_PHYDS4S_h	**Lycopene**
R_ZDS_ADDED	R_ZDS_h	all-*trans*-Neurosporene
R_Ctrl_ADDED	R_NOR_h	Lycopene
R_LCYB_ADDED	R_LYCBC1_c	gamma-Carotene
R_LCYB2_ADDED	R_LYCBC2_h	beta-Carotene
R_LUT1_2_1__ADDED		**beta-Cryptoxanthin**
R_LUT1_2_2__ADDED		**(beta-Cryptoxanthin) (Zeaxanthin)**
	R_BCAROXR_h	**Zeaxanthin**
R_ZEP_1__ADDED	R_ZXANEPOX_h; R_ANTHXANDE_h	(Zeaxanthin, Antheraxanthin)
R_ZEP_2__ADDED	R_ZXANEPOX_h R_ANTHXANDE_h	(Zeaxanthin, Antheraxanthin)
R_ZEP_3__ADDED	R_ANTHXANEPOX_h R_VIOXANDE_h	(Antheraxanthin, Violaxanthin)
R_ZEP_4__ADDED	R_ANTHXANEPOX_h R_VIOXANDE_h	(Antheraxanthin, Violaxanthin)
R_VDL_ADDED	R_ANTHXANDE_h; R_ZXANEPOX_h	(Antheraxanthin, Zeaxanthin)
R_VDL_2__ADDED	R_ANTHXANEPOX_h R_VIOXANDE_h	(Violaxanthin, Antheraxanthin)
R_DIADINO_ADDED		**(Violaxanthin, diadinoxanthin)**
R_FUCO_ADDED		**(Diadinoxanthin, fucoxanthin)**
R_FUCO_ADDED_drain	R_FXD	Drain for fucoxanthin
R_NEOX_ADDED	R_NSY_h	(Violaxanthin, neoxanthin)
	R_FXANS_h	**(Neoxanthin, fucoxanthin)**
	R_DIADINX_h	**(Neoxanthin, diadinoxanthin)**
R_CISNEOX_ADDED		**(Neoxanthin, 9′ -*cis*-Neoxanthin)**
R_DIATO_ADDED	R_DIADINXDE_h	(Diadinoxanthin, diatoxanthin)
R_DIADINO2_ADDED	R_DIATOXEPOX_h	(diatoxanthin, Diadinoxanthin)

The modified model was uploaded to OptFlux software [9] to perform *in silico* metabolic engineering tasks. Simulations using this altered model were made regarding inorganic carbon (CO_2 and HCO_3^-) and nitrogen supplies (urea and NO_3^-). Next, another genome-scale metabolic model for *P. tricornutum* published by Levering et al. [7], which includes the reactions that complete the MEP pathway for the synthesis of fucoxanthin was used for phenotype simulation and strain optimization for fucoxanthin production as before.

There are some slight, but not unimportant, differences in some MEP pathway reactions of these two genome-scale models shown in Table 1, besides the difference in the number of cellular compartments, i.e., five in the former one and six in the latter. In the first model [6], fucoxanthin is synthesized from diadinoxanthin as proposed by Coesel et al. in 2008 [10], whereas in the other model [7] fucoxanthin is synthesized from neoxanthin as predicted by Mikame and Hosokawa [2] in 2013.

One can think of two forms to approach fucoxanthin production, i.e., enhancing biomass production and therefore also fucoxanthin, guaranteeing that its production is coupled to biomass, or eliciting cells to drive their metabolism towards fucoxanthin production. Even better is to combine both approaches. Thus, phenotype simulations are done mainly to achieve biomass growth, by using optimized culture media, with combinations of different carbon and nitrogen sources. On the other hand, strain optimization algorithms are used to reach possible ways to enhance fucoxanthin biosynthesis through genetic modifications. Fucoxanthin is not a compound excreted by the cell. Therefore, it is intrinsically related to growth, since it takes part of the photosynthetic apparatus essential for growth of photosynthetic cells. Indeed, in this last model tested [7], fucoxanthin is included in the biomass equation. The presence of fucoxanthin in fucoxanthin-chlorophyll *a/c* binding proteins (FCP) agrees with the light harvesting function of this pigment and its importance in biomass construction. Thus, one can tightly infer that fucoxanthin is closely enhanced with biomass growth.

To perform the phenotype simulation and strain optimization tasks, parameters for biomass and fucoxanthin production were established through a simulation with the original conditions of the published models regarding reactions to be constrained thereafter (Table 2). Strain optimization runs were done using Biomass-product coupled yield (BPCY) as the objective function and Minimization of metabolic adjustment (MOMA) [11] as phenotype prediction method.

Table 2. Original constraints used for simulations - carbon/nitrogen sources.

Compound	Flux [mmol gDW^{-1} h^{-1}]*
CO_2	1.3
HCO_3^-	0.65
NO_3^-	308.94
Urea	0
NH3/NH4$^+$	0

*mmol: milimol, gDW: grams of dry weight, h: hour

3   Results and Discussion

The results for growth simulations were almost identical for both models and are provided in Table 3. The biomass flux increased substantially with doubled supply of inorganic carbon, i.e., 67%, 33%, and 100% for CO_2, HCO_3^- and CO_2 plus HCO_3^-, respectively, and 77% for doubled urea supply. Urea metabolism releases NH_3/NH_4^+ and CO_2 being this surplus of CO_2 thought to be responsible for biomass increase. Interestingly, when urea supply is increased, NH_3/NH_4^+ had to be drained to obtain an increase in biomass. Glibert et al., [12] and references therein report growth inhibition and even toxic effect of NH_4^+ in the cells. When NH_3/NH_4^+ is supplied without urea, *P. tricornutum* biomass production practically remains unchanged. NO_3^- seems to exhibit a limited uptake around 0.15 mmol gDW^{-1} h^{-1} [6] and 0.21 mmol gDW^{-1} h^{-1} [7], and above those limits biomass flux did not change as for lower values biomass decreased.

According to the predictions of both models studied, it is possible to enhance biomass production by supplying N_2 and inorganic carbon, being the best results obtained through a combination of CO_2 and HCO_3^- or, regarding nitrogen, supplementing the culture medium with urea. The increment in biomass flux under constraints for carbon and nitrogen was numerically equivalent in both models tested. Unlike metazoans that utilize urea cycle to excrete N_2, diatoms utilize this pathway to synthesize nitrogen compounds [13]. Preliminary *in vitro* tests supplying urea (160 mM) to *P. tricornutum* cultures increased cell density by the fourth and seventh days at 6-fold and 2.7-fold higher, respectively (data not shown).

Table 3. Simulations performed on OptFlux for biomass maximization under specific constraints (constraints are detailed in Table 2).

Constraints [mmol gDW^{-1} h^{-1}]*	Biomass flux [mmol gDW^{-1} h^{-1}]*	
	Kim et al. [6]	Levering et al. [7]
Parameter for biomass growth	0.05013	0.04109
CO_2 – Double supply (2.6)	0.08355 (> 67%)	0.06848 (> 67%)
HCO_3^- – Double supply (1.3)	0.06684 (> 33%)	0.05478 (> 33%)
CO_2 e HCO_3^- – Double supply (2.6 and 1.3 respectively)	0.10027 (> 100%)	0.08850 (> 100%)
Urea - Double supply (1.5)	0.08870 (> 77%)	0.07270 (> 77%)
NO_3 (0.15 or 0.21)	0.05013 (No change)	0.04109 (No change)
NO_3 (0.5) (> 0.15 or 0.21)	0.05013 (No change)	0.04109 (No change)
NO_3 (0.1) (< 0.15 or 0.21)	0.03301 (–34%)	0.04087 (–0.5%)

*mmol: milimol, gDW: grams of dry weight, h: hour

Strain optimization methods for fucoxanthin production were applied also on OptFlux for these models [6, 7] using BPCY as the objective function (Biomass flux x fucoxanthin drain)/CO_2 exchange). The results for biomass and fucoxanthin flux are provided in Table 4. In both models, fucoxanthin flux in wild types returned very low

Table 4. Optimizations performed on OptFlux for fucoxanthin maximization under specific constraints (constraints are detailed in Table 2).

Model	Environmental conditions (*constraints*) [mmol gMS^{-1} h^{-1}]*	Knockout (reactions)	Biomass parameter (wild type) [mmol gMS^{-1} h^{-1}]*	Biomass coupled to fucoxanthin production optimization [mmol gMS^{-1} h^{-1}]*	Fucoxanthin parameter (wild type) [mmol gMS^{-1} h^{-1}]*	Fucoxanthin production optimized [mmol gMS^{-1} h^{-1}]*
Kim et al. [6]	original conditions (Table 2)	R_TPI_P	0.05013	0.02299 (<55%)	0	0.01433
	CO_2 – Double supply (2.6)	R_NIR_P R_PPDK_P	0.08355	0.04342 (<48%)	0	0.03083
	HCO_3^- – Double supply (1.3)	R_NIR_P R_GS_P	0.06684	0.04191 (<37%)	0	0.02126
		R_PPDK_P R_GOGAT_FD_P R_NIR_P	0.06684	0.02974 (<55%)	0	0.03177
	CO_2 e HCO_3^- – Double supply (2.6 and 1.3)	R_NIR_P R_GAPDH	0.10027	0.07279 (<27%)	0	0.01887
	Urea - (1.5)	R_TPI_P	0.08870	0.04430 (<50%)	0	0.02636

(*continued*)

Table 4. (*continued*)

Model	Environmental conditions (*constraints*) [mmol gMS^{-1} h^{-1}]*	*Knockout* (reactions)	Biomass parameter (wild type) [mmol gMS^{-1} h^{-1}]*	Biomass coupled to fucoxanthin production optimization [mmol gMS^{-1} h^{-1}]*	Fucoxanthin parameter (wild type) [mmol gMS^{-1} h^{-1}]*	Fucoxanthin production optimized [mmol gMS^{-1} h^{-1}]*
Levering et al. [7]	original conditions (Table 2)	R_NTRIR_h	**0.04109**	0.01981 (<52%)	**0.00046**	0.01045 (> 23-fold)
	CO_2 – Double supply (2.6)	R_NTRIR_h	**0.06848**	0.03442 (<50%)	**0.00076**	0.01828 (> 24-fold)
	HCO_3^- – Double supply (1.3)	R_GLNA_h	**0.05478**	0.03983 (<27%)	**0.00061**	0.00909 (> 15-fold)
	CO_2 e HCO_3^- – Double supply (2.6 and 1.3)	R_GLTS_h	**0.08218**	0.05475 (<33%)	**0.00091**	0.01559 (> 17-fold)
	Urea - (1.5)	R_GLTS_h	**0.07269**	0.05805 (<20%)	**0.00081**	0.00833 (> 10-fold)

*mmol: milimol, gDW: grams of dry weight, h: hour

values when compared to the modified strains. Indeed, for the first model [6] no fucoxanthin production was observed without optimization. Since optimizations return numerous possible knockout solutions, here we try to highlight some of them for future *in vitro* assays. Disregarding transport reactions, knockouts for fucoxanthin production in both models were for nitrite reductase (NR) (EC 1.7.7.1), glutamine synthase (GS) (EC 6.3.1.2), and glutamate synthase (GOGAT) (EC 1.4.7.1); all them occur in plastids. For the model of Kim et al. [6], knockouts for pyruvate-phosphate dikinase (PPDK)(EC 2.7.9.1), glyceraldehyde-3-phosphate dehydrogenase (G3P) (EC 1.2.1.12) and triose-phosphate isomerase (TPI) (EC 5.3.1.1) were also good solutions. NR as well as GS and GOGAT are involved in amino acid biosynthesis and knockout of these reactions can possibly leave more reductants free for fucoxanthin biosynthesis. Knockout of PPDK prevents phosphoenolpyruvate (PEP) formation remaining more pyruvate free for the MEP pathway. G3P and TPI in turn prevent G3P catabolism remaining also more G3P free for MEP. These are important precursors in fucoxanthin biosynthesis.

Choi and coworkers [14], using an *in silico* tool called flux scanning based on enforced objective flux (FSEOF), determined increased fluxes in lycopene production, a carotenoid produced via MEP pathway in *Escherichia coli*. The authors [14] reported that fluxes in the upper glycolytic pathway should increase, suggesting G3P availability as being important for lycopene production. Haimovich-Dayan et al., 2013 [15] lowered PPDK activity in *P. tricornutum* transformant cells and found an increase in carbohydrates and lipid levels, contrasting to a decrease in protein amounts and an apparent increase in non-photochemical quenching (NPQ) as a possible consequence of an activation of the xanthophyll cycle and reaction centers. However, how PPDK knockdown affects fucoxanthin biosynthesis has to be experimentally validated yet.

4 Conclusions

For both models tested, *P. tricornutum* biomass growth was predicted to increase in all sources tested, being the best results obtained with a combination of CO_2 plus HCO_3^- regarding carbon sources and with urea regarding nitrogen source. Many of the knockout reactions brought on as a result of optimizations for fucoxanthin production are involved in the metabolism of pyruvate, glutamine/glutamate and nitrogen assimilation and some of the solutions still have to be experimentally validated.

Acknowledgements. This work was supported by a grant from the National Council for Scientific and Technological Development (CNPq n° 490383/2013-0). The research fellowship from CNPq (grant n° 307099/2015-6) on behalf of M. Maraschin is acknowledged.

References

1. Kawee-ai, A., Kuntiya, A., Kim, S.M.: Anticholinesterase and antioxidant activities of fucoxanthin purified from the microalga Phaeodactylum tricornutum. Nat. Prod. Commun. **8**(10), 1381–1386 (2013)
2. Mikami, K., Hosokawa, M.: Biosynthetic pathway and health benefits of fucoxanthin, an algae-specific xanthophyll in brown seaweeds. Int. J. Mol. Sci. **14**(7), 13763–13781 (2013)
3. Kim, S.M., Jung, Y.J., Kwon, O.N., et al.: A potential commercial source of fucoxanthin extracted from the microalga Phaeodactylum tricornutum. Appl. Biochem. Biotechnol. **166**(7), 1843–1855 (2012)
4. Bowler, C., Allen, A.E., Badger, J.H., et al.: The Phaeodactylum genome reveals the evolutionary history of diatom genomes. Nature **456**(7219), 239–244 (2008)
5. Fabris, M., Matthijs, M., Rombauts, S., et al.: The metabolic blueprint of Phaeodactylum tricornutum reveals a eukaryotic Entner-Doudoroff glycolytic pathway. Plant J. **70**(6), 1004–1014 (2012)
6. Kim, J., Fabris, M., Baart, G., et al.: Flux balance analysis of primary metabolism in the diatom Phaeodactylum tricornutum. Plant J. **85**(1), 161–176 (2016)
7. Levering, J., Broddrick, J., et al.: Genome-scale model reveals metabolic basis of biomass partitioning in a model diatom. PLoS One **11**(5), e0155038 (2016)
8. McClure, D.D., Audrey, L., Blandine, G., et al.: An investigation into the effect of culture conditions on fucoxanthin production using the marine microalgae *Phaeodactylum tricornutum*. Algal Res. **29**, 41–48 (2018)
9. Rocha, I., Maia, P., et al.: Optflux: an open-source software platform for *in silico* metabolic engineering. BMC Sys. Biol. **4**(1), 45 (2010)
10. Coesel, S., Oborník, M., Varela, J., et al.: Evolutionary origins and functions of the carotenoid biosynthetic pathway in marine diatoms. PLoS ONE **3**(8), e2896 (2008)
11. Segre, D., Vitkup, D., Church, G.M.: Analysis of optimality in natural and perturbed metabolic networks. Proc. Natl. Acad. Sci. U.S.A. **99**(23), 15112–15117 (2002)
12. Glibert, P.M., Wilkerson, F.P., Dugdale, R.C., et al.: Pluses and minuses of ammonium and nitrate uptake and assimilation by phytoplankton and implications for productivity and community composition, with emphasis on nitrogen-enriched conditions. Limnol. Oceanogr. **61**, 165–197 (2016)
13. Allen, A.E., Dupont, C.L., et al.: Evolution and metabolic significance of the urea cycle in photosynthetic diatoms. Nature **473**(7346), 203–207 (2011)
14. Choi, H.S., Lee, S.Y., et al.: *In silico i*dentification of gene amplification targets for improvement of lycopene production. Appl. Environm. Microbiol. **76**(10), 3097–3105 (2011)
15. Haimovich-Dayan, M., Garfinkel, N., et al.: The role of C_4 metabolism in the marine diatom *Phaeodactylum tricornutum*. New Phytol. **197**(1), 177–185 (2013)

EvoPPI: A Web Application to Compare Protein-Protein Interactions (PPIs) from Different Databases and Species

Noé Vázquez[1,2], Sara Rocha[3,4], Hugo López-Fernández[1,2,3,4,5(✉)],
André Torres[3,4], Rui Camacho[6], Florentino Fdez-Riverola[1,2,5],
Jorge Vieira[3,4], Cristina P. Vieira[3,4], and Miguel Reboiro-Jato[1,2,5]

[1] ESEI - Escuela Superior de Ingeniería Informática, Universidad de Vigo,
Edificio Politécnico, Campus Universitario As Lagoas s/n, 32004 Ourense, Spain
nvazquezg@gmail.com,
{hlfernandez, riverola, mrjato}@uvigo.es
[2] Centro de Investigaciones Biomédicas
(Centro Singular de Investigación de Galicia), Vigo, Spain
[3] Instituto de Investigação e Inovação em Saúde (I3S),
Universidade do Porto, Rua Alfredo Allen, 208, 4200-135 Porto, Portugal
sara.rocha@i3s.up.pt,
{andre.torres, jbvieira, cgvieira}@ibmc.up.pt
[4] Instituto de Biologia Molecular e Celular (IBMC),
Rua Alfredo Allen, 208, 4200-135 Porto, Portugal
[5] SING Research Group, Galicia Sur Health Research Institute (IIS Galicia Sur),
SERGAS-UVIGO, Vigo, Spain
[6] LIAAD, DEI, Faculdade de Engenharia,
Universidade do Porto, Porto, Portugal
rcamacho@fe.up.pt

Abstract. Biological processes are mediated by protein-protein interactions (PPI) that have been studied using different methodologies, and organized as centralized repositories - PPI databases. The data stored in the different PPI databases only overlaps partially. Moreover, some of the repositories are dedicated to a species or subset of species, not all have the same functionalities, or store data in the same format, making comparisons between different databases difficult to perform. Therefore, here we present EvoPPI (http://evoppi.i3s.up.pt), an open source web application tool that allows users to compare the protein interactions reported in two different interactomes. When interactomes belong to different species, a versatile BLAST search approach is used to identify orthologous/paralogous genes, which to our knowledge is a unique feature of EvoPPI.

Keywords: Protein-protein interactions databases · Inter-specific comparisons
Graphical view

© Springer Nature Switzerland AG 2019
F. Fdez-Riverola et al. (Eds.): PACBB 2018, AISC 803, pp. 149–156, 2019.
https://doi.org/10.1007/978-3-319-98702-6_18

1   Introduction

Most proteins are functional only when present as homo/hetero complexes. Finding which proteins interact thus provides opportunities for the exploration of protein functions [1]. Moreover, aberrant protein-protein interactions (PPIs) are observed in multiple aggregation-related diseases, such as Creutzfeldt–Jakob, Alzheimer's, Parkinson's, cancer and polyglutamine diseases [2, 3]. PPI networks comparison in controls and patients may thus give insight into the basis of the disease, and suggest possible therapeutic targets.

Many experimental and computational methods based on single or high-throughput screens have been developed for detecting PPIs [4]. All PPI detection methods have advantages and disadvantages. For instance, the three most used experimental high-throughput measurement techniques (tandem affinity purification, two-hybrid assays, and mass spectrometry) have very high rates of false positive interactions, sometimes up to 50% [1]. It should be noted that, even when using high-throughput methods there is a significant chance of not being able to recover all interactions. On the other hand, methods such as X-ray crystallography that are unlikely to produce false positives are difficult to scale up, and thus cannot be used to address large numbers of PPIs. Computational methods, such as docking, text mining, interolog mapping, machine learning and so forth [5] can detect thousands of PPIs at a very low cost and in much less time than experimental methods. Nevertheless, when using such methods, the accuracy is always an issue since they are not based on experimental data but rather on predictions, although they are very useful to understand which interactions may be missing in the available experimental datasets.

PPI datasets for different species, using multiple detection approaches, are publicly available in databases, such as BioGRID, CCSB, DroID, FlyBase, HIPPIE, HitPredict, HomoMINT, Instruct, Interactome3D, mentha, MINT, PINA, PRIN, TAIR, or APID [1, 6, 7]. Although there is some degree of overlap between the different databases, each presents a unique set of information, as a consequence of the usage of a different rational for database building. The PPI datasets can be downloaded, but the interactions are reported in different formats making any comparison across databases very difficult. For instance, BioGRID, MINT and CCSB use gene identifiers, UniProt numbers, and gene names, respectively, to report interactions.

Some databases (BioGRID, MINT) have a label attached to each PPI that states the source of the reported interaction, while others do not, and some (BioGRID, HIPPIE, and MINT) databases are human-curated. Moreover, although protein complexes may be conserved across different species, functionally equivalent proteins may have different names in different species, making any kind of comparison very difficult. Given the known advantages and disadvantages of every method and database, it is clear, for instance, that interactions reported in more than one independent study, or even in different species should be considered more likely to be true interactions than those reported in a single study, using high-throughput methods. Moreover, as stated above, the comparison of different types of interactomes such as those from controls and patients is of interest.

Here we present EvoPPI (http://evoppi.i3s.up.pt), a web application that can be used to effortlessly compare PPI datasets from different databases and species. For the latter case, a BLAST-based approach is used. By adjusting parameters such as the number of descriptions to report, expect value, percentage of identity and minimum size of the aligned block, users have the ability to decide which proteins are functionally equivalent in different species. This feature is especially useful when comparing model species such as *Drosophila melanogaster* and *Homo sapiens*, since in the lineage leading to the latter species two rounds of whole genome duplication have occurred [8], meaning that most genes in *Drosophila* have multiple paralogs in humans. EvoPPI has some unique features such as the use of a versatile BLAST approach for defining orthologous/paralogous genes (that allows, for instance, to predict PPI networks in species where such data does not exist using a comparative approach), and the use of colour codes for an easy visualization of the differences between datasets (see Supplementary Table 1 available at https://doi.org/10.5281/zenodo.1204854 for a comparison with similar tools).

2 Materials and Methods

EvoPPI is comprised of two different applications that act as the front-end and the back-end components, respectively. The front-end application is a web application implemented using the Angular v5.2 framework[1] in combination with the Angular Material v5.2 library[2], for a richer user interface. The back-end application has been implemented using the Java EE 7 platform[3]. This application provides a RESTful API [9] with resources to access data and to request PPIs operations. Communication between front-end and back-end applications is done using AJAX (Asynchronous JavaScript And XML) and JSON (JavaScript Object Notation) for data encoding.

EvoPPI relies on BLAST to perform sequence alignment between the protein sequences of different species, in order to identify orthologous/paralogous genes. This identification is needed to make possible the comparison of interactomes belonging to different species. In order to avoid installation and configuration issues, a Docker[4] container has been created with a BLAST v2.6.0 installation. This container is invoked from the back-end application using the docker-java v3.0.13 library[5].

Figure 1 represents the general architecture and deployment of EvoPPI, including the components described above. EvoPPI is currently running in a WildFly v10.1.0 application server[6] and uses a MySQL v5.7 database management system[7] to store the information.

---

[1] https://angular.io/.

[2] https://material.angular.io/.

[3] https://www.oracle.com/es/java/technologies/java-ee.html.

[4] https://www.docker.com/.

[5] https://github.com/docker-java/docker-java.

[6] http://wildfly.org/.

[7] https://www.mysql.com/.

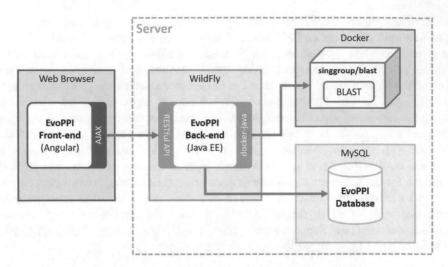

Fig. 1. General architecture and components of EvoPPI.

EvoPPI is freely available at http://evoppi.i3s.up.pt. It is an open source software distributed under a GPLv3 license. The source code of the front-end application is publicly available at https://github.com/sing-group/evoppi-frontend, while the source code of the back-end application is available at https://github.com/sing-group/evoppi-backend. Finally, the Docker container with the BLAST installation is available at https://hub.docker.com/r/singgroup/evoppi-blast/.

Intra-species comparisons are performed using two or more files where the protein interactions are listed using GeneBank identifiers, and another one describing the known synonyms for each gene (a file prepared from the GenBank file for the genome of the species being considered). When the user types the name of a protein, the corresponding gene identifier is searched and the corresponding interactions, listed in the selected databases, shown. When performing inter-species comparisons, two additional files are used, namely two protein FASTA files with all reported proteins for the query and target species (prepared from GenBank files for the genomes of the species being considered, and having as the header GeneBank identifiers). The homolog of the protein of interest is identified using *blastp* and the parameters chosen by the user. Then, two independent networks are obtained for the query and target species. Finally, the correspondences between the nodes of the two networks being compared are established from the results of the *blastp* search using the interacting proteins of the query species as the query. We advise users to perform the inter-species comparisons using different parameters for the BLAST search and associated filters, and compare the results, since it is impossible to predict the best default values for every case.

3 Results and Discussion

The case study is Ataxin-2 (ATXN2), a polyQ protein encoded by the *ataxin-2* gene located in human chromosome 12. When the polyQ stretch expands above 31 glutamines, individuals show Spinocerebellar Ataxia 2 (SCA2) [10], an autosomal dominant disease. In order to understand why this happens it is relevant to identify the interactions established between the ATXN2 and other proteins. Nevertheless, there are different databases representing PPI, where different methodologies have been used (see Introduction). Therefore, it is of interest to compare the different PPI reported in databases that use different approaches. Here we use the easy-to-use "Compare same species" EvoPPI panel to compare the interactions reported for ATXN2 in HomoMINT (that uses the inparanoid algorithm) and Interactome3D (homology modelling pipeline that combines Intact20, MINT21, DIP44, MPIDB45, MatrixDb46, InnateDb47, BioGRID48 and BIND databases). Although EvoPPI uses GenBank identifiers as the main identifier for the genes, it also stores other alternative names. When a user starts to type a gene name, EvoPPI looks for that text in the gene identifier and alternative names, to show a list of genes from which the user can select the query protein. In addition to these parameters, the Interaction level parameter can be used to select the maximum degree of distance of the retrieved interactions.

Using interaction level 1, by comparing the *Homo sapiens* at the HomoMINT and at the Interactome3D databases, 14 interactions were obtained (13 are only found in HomoMINT and one is only found in Interactome3D). Results are available as a table that lists the interactions, including the gene identifiers and names, and the interaction degree in each interactome (that can be exported in CSV format) and as an undirected graph where nodes are genes and edges are interactions (Fig. 2). In the latter, a colour code is used (a blue, black and red connector line indicates that the interaction is reported only in the reference interactome, only in the target interactome or in both, respectively). In both views, proteins can be clicked to view detailed information, including the gene identifier, alternative names, and protein sequence(s). The protein sequence(s), associated with the query and interacting proteins can be downloaded in FASTA format. This allows users to perform additional operations using other applications (for instance, predict the 3D structure of the target and interacting protein as well as determining the docking surface). A permalink to the results page can also be created allowing the user to return later on to the results. When interaction level 2 is used, 726 interactions are found. In this case, a graph view is not available (the limit for the number of interactions shown as a graph, above interaction level 1, is 100). These results show that although ATXN2 establishes 14 interactions with other proteins, those in turn establish a large number of interactions with other proteins. Since such a result is observed in most proteins, the highest interaction level that is allowed is 3. In the case of ATXN2, 4757 interactions are retrieved when using interaction level 3.

Since an evolutionary comparative approach can also be informative, we address how many of the reported ATXN2 interactions reported for *H. sapiens* in the HomoMINT database are also reported for *D. melanogaster* in the Interactome3D database. For this, we use the EvoPPI "Compare species" panel. Since this requires the identification of orthologous/paralogous genes in distantly related species, a BLAST approach

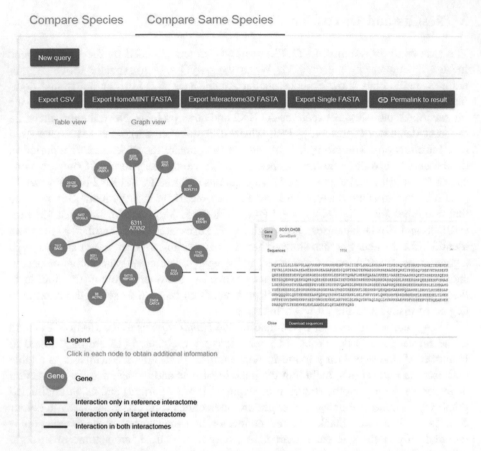

Fig. 2. Screenshot of EvoPPI "Compare same species" panel showing the result for ATXN2 when comparing HomoMINT and Interactome3D databases. The broken line indicates the information that is shown when clicking on top of the SCG1 protein.

is used. It should be noted, that depending on the proteins and species being considered, users might want to use different criteria for orthologous/paralogous gene identification. Therefore, in EvoPPI the users indicate the expect value to be used in the *blastp* search, the minimum length of the aligned block to be considered, the number of descriptions (hits), and minimum identity over the aligned fragment being considered. Given the large evolutionary time between *D. melanogaster* and *H. sapiens*, we use an expect value of 0.05, minimum length of alignment block of 50, and minimum identity of 40%. Although "Compare same species" queries are performed within a few seconds, the execution of different species queries may take several minutes or even hours, due to the sequence alignment step. Taking this into account, EvoPPI was designed to execute the queries asynchronously, so that users can launch a query, leave the application and return later to check the execution status. To do so, EvoPPI generates a unique URL for each query that users can use to return at any time to the query result. The results show that none of the 13 interactions present in HomoMINT have been reported in the

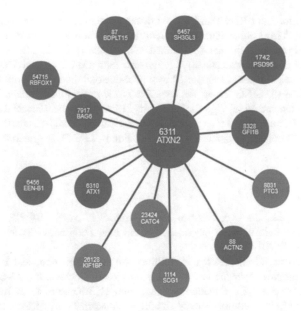

Fig. 3. PPI network generated with the EvoPPI "Compare different species" option, showing the result for ATXN2 when comparing *H. sapiens* HomoMINT and *D. melanogaster* Interactome3D databases. A green background indicates that an orthologous/paralogous gene has been identified in *D. melanogaster* (see text for details).

D. melanogaster interactome3D database. Nevertheless, the orthologous/paralogous of 10 genes (those with a green background) have been identified in *D. melanogaster* (Fig. 3). By clicking on top of any protein on the graph view, details are given regarding the BLAST operations, including the geneID of the genes identified as orthologous/ paralogous, as well as their protein sequence(s).

4   Conclusions

EvoPPI is an open source web application tool that allows users to compare the interactions of a protein in two different interactomes, even if the interactomes belong to different species. The latter comparisons are achieved by using a versatile BLAST search approach which to our knowledge, is a unique feature of EvoPPI, not found in other tools that can only identify orthologs in different species when they have the same gene name. The comparison of different datasets obtained with different methodologies, and even for different species can, in principle, be used as a tool to reduce the false positive rate. The current development of EvoPPI is focused in two main lines to increase its functionalities: (*i*) add user accounts, so that users can keep and manage their query results, and (*ii*) create a management interface, to allow users to feed the EvoPPI database with new interactomes and species.

Acknowledgements. Sara Rocha and this work are supported by the project Norte-01-0145-FEDER-000008 - Porto Neurosciences and Neurologic Disease Research Initiative at I3S, supported by Norte Portugal Regional Operational Programme (NORTE 2020), under the POR-TUGAL 2020 Partnership Agreement, through the European Regional Development Fund (FEDER). H. López-Fernández is supported by a post-doctoral fellowship from Xunta de Galicia (ED481B 2016/068-0). SING group thanks CITI (*Centro de Investigación, Transferencia e Innovación*) from University of Vigo for hosting its IT infrastructure. Financial support from the Xunta de Galicia (Centro singular de investigación de Galicia accreditation 2016-2019) and the European Union (European Regional Development Fund - ERDF), is gratefully acknowledged.

References

1. Alanis-Lobato, G., Andrade-Navarro, M.A., Schaefer, M.H.: HIPPIE v2.0: enhancing meaningfulness and reliability of protein–protein interaction networks. Nucleic Acids Res. **45**, D408–D414 (2017)
2. Chiti, F., Dobson, C.M.: Protein misfolding, amyloid formation, and human disease: a summary of progress over the last decade. Annu. Rev. Biochem. **86**, 27–68 (2017)
3. Cescatti, M., Saverioni, D., Capellari, S., Tagliavini, F., Kitamoto, T., Ironside, J., Giese, A., Parchi, P.: Analysis of conformational stability of abnormal prion protein aggregates across the spectrum of Creutzfeldt-Jakob disease prions. J. Virol. **90**, 6244–6254 (2016)
4. Peng, X., Wang, J., Peng, W., Wu, F.-X., Pan, Y.: Protein–protein interactions: detection, reliability assessment and applications. Brief. Bioinform. **18**, 798–819 (2017)
5. Folador, E., de Oliveira Junior, A., Tiwari, S., Jamal, S., Ferreira, R., Barh, D., Ghosh, P., Silva, A., Azevedo, C.: In silico protein-protein interactions: avoiding data and method biases over sensitivity and specificity. Curr. Protein Pept. Sci. **16**, 689–700 (2015)
6. Bader, G.D., Cary, M.P., Sander, C.: Pathguide: a pathway resource list. Nucleic Acids Res. **34**, D504–D506 (2006)
7. Alonso-López, D., Gutiérrez, M.A., Lopes, K.P., Prieto, C., Santamaría, R., De Las Rivas, J.: APID interactomes: providing proteome-based interactomes with controlled quality for multiple species and derived networks. Nucleic Acids Res. **44**, W529–W535 (2016)
8. Mendivil Ramos, O., Ferrier, D.E.K.: Mechanisms of Gene Duplication and Translocation and Progress towards Understanding Their Relative Contributions to Animal Genome Evolution. https://www.hindawi.com/journals/ijeb/2012/846421/
9. Fielding, R.T.: Architectural Styles and the Design of Network-Based Software Architectures (2000)
10. Ng, H., Pulst, S.-M., Huynh, D.P.: Ataxin-2 mediated cell death is dependent on domains downstream of the polyQ repeat. Exp. Neurol. **208**, 207–215 (2007)

A Review on Metabolomics Data Analysis for Cancer Applications

Sara Cardoso[✉], Delora Baptista, Rebeca Santos, and Miguel Rocha

CEB - Centre Biological Engineering, University of Minho,
Campus of Gualtar, Braga, Portugal
saracardoso501@gmail.com

Abstract. Cancer cells undergo metabolic changes that contribute to tumorigenesis, which can be determined using metabolomics data produced by techniques such as nuclear magnetic resonance and mass spectroscopy, and analyzed through statistical and machine learning methods. Since these data represent well the metabolic phenotype of these cells, they are very relevant in cancer research, to better understand tumour cells metabolism and help in efforts of biomarker and drug target discovery. This mini-review focuses on data analysis methods that are commonly used to extract knowledge from cancer metabolomics data, such as univariate analysis and supervised and unsupervised multivariate data analysis, including clustering and machine learning.

Keywords: Cancer · Metabolomics · NMR · Mass spectrometry
Machine learning · Chemometrics

1 Introduction

Cancer, a label applied to a variety of diseases featuring excessive cell proliferation, is driven by changes at the genomic level, which define a distinct metabolic profile that supports the tumorigenic process. A common alteration, usually referred to as the Warburg effect [1], is the observation that cancer cells resort to glycolysis with subsequent lactate fermentation to produce energy, even under aerobic conditions. Many other metabolic changes have since been documented, and a recent review has identified six cancer metabolism hallmarks [2].

These changes in intracellular, extracellular, and circulating metabolites can be assessed by applying one of two approaches. Targeted studies focus on a selected subset of known metabolites, while untargeted studies attempt to profile the metabolome in a non-predefined manner. The metabolomics data can be obtained using techniques such as Nuclear Magnetic Resonance (NMR) spectroscopy, and Mass Spectroscopy, normally coupled to Gas or Liquid Chromatography (GC/LC-MS).

NMR has been extensively used for several purposes in cancer studies, such as the distinction between tumor and normal samples [3], prediction of patient survival [4] and tumor recurrence [5], and monitoring tumor drug response [6].

© Springer Nature Switzerland AG 2019
F. Fdez-Riverola et al. (Eds.): PACBB 2018, AISC 803, pp. 157–165, 2019.
https://doi.org/10.1007/978-3-319-98702-6_19

Table 1. Data analysis methods used in a selection of recent cancer NMR and MS studies.

Study description	Technique	Analysis	Ref.
Discrimination between advanced colorectal cancer patients and controls	NMR	PCA; HC; OPLS-DA; univariate tests; CA	[13]
Pancreatic adenocarcinoma vs healthy; biomarker discovery	NMR	OPLS-DA; MW	[4]
Identification of metastatic colorectal cancer samples and survival prediction	NMR	PLS; SVM	[14]
Metabolic profiles of breast cancer cell lines undergoing treatment	NMR	HC; ANOVA; HSD	[15]
Discrimination of gastric cancer from benign disease based on urine samples	NMR	PLS-DA; OPLS-DA; MW	[16]
Discrimination between hepatocellular carcinoma samples and controls	NMR	OPLS-DA; ANOVA	[17]
Discrimination of metastatic breast cancer patients; risk of recurrence	NMR	PCA; RF; MW	[5]
Tamoxifen-resistant vs CK α downregulated breast cancer cells	NMR	PCA; PLS-DA; TT; ANOVA	[18]
Metabolic profiling of urine samples for renal cell carcinoma diagnosis	NMR	PCA; PLS-DA; TT; MW	[19]
Response of glioblastoma cell lines treated with 5-lipoxygenase inhibitors	NMR	PCA	[20]
Discrimination between basal cell carcinoma sample and normal controls	NMR	OPLS-DA; TT	[21]
Metabolomics of seminal plasma for prostate cancer diagnosis	NMR	PCA; PLS-DA; MW	[22]
Metabolic profiling of glioma cell lines with different degrees of malignancy	NMR	PCA; PLS-DA; OPLS-DA; CA; TT	[23]
Metabolic characterization of breast cancer cells experiencing hypoxia	NMR	PCA; SVM; TT; FC	[24]
Studying the effect of soy isoflavones on breast cancer cell lines	NMR	PCA; HC	[25]
Tissue metabolic profiling of human gastric cancer	NMR	PCA; PLS-DA; OPLS-DA; TT	[3]
Serum and urine samples to diagnose cancer cachexia	NMR	OPLS-DA; RF; TT; ANOVA-HSD; Chi^2; CA	[26]
Temporal characterization of serum metabolite signatures in lung cancer patients undergoing treatment	NMR; MS	ANOVA; PCA; OPLS-DA	[7]
Metabolomic Changes in Blood Samples of Lung Cancer Patients	MS	ANOVA	[27]

(*continued*)

Table 1. (*continued*)

Study description	Technique	Analysis	Ref.
Lipid metabolism, ER stress and induction of an acute inflammatory response as a target in renal cell carcinoma	MS	PLS; PCA; TT; ANOVA	[28]
Identification of hepatocellular carcinoma biomarkers	MS	ANOVA; PLS-DA; OPLS-DA	[12]
Identification of hepatocellular carcinoma biomarkers	MS	ANOVA	[29]
Distinguishing metabolic profiles of OVCAR-3 spheroid-derived cells	MS	TT; PCA; HC	[11]
Metabolomic study of apoptosis in an *in vitro* leukemia model	MS	ANOVA; TT; PCA	[30]
Characterization of lipid metabolism in normal tissues vs colorectal cancer	MS	MW; KW; ANOVA	[31]
Detection of colorectal cancer using targeted serum metabolic profiling	MS	MW; PLS-DA	[10]
Metabolic responses in an ovarian cancer stem cell line	MS	ANOVA; TT; PCA	[32]
Identification of diagnostic/prognostic markers for lung cancer	MS	RF; TT	[9]
Association of chemotherapy response in breast cancer with fatty acids	MS	ANOVA; PLS-DA; TT	[33]
Biomarker identification for hepatocellular carcinoma	MS	TT; MW	[34]

On the other hand, applications of MS in cancer research include the characterization of metabolite signatures in lung cancer patients undergoing treatment [7], and several cases of metabolic profiling to find diagnostic/prognostic biomarkers of tumors like lung, colorectal, ovarian and hepatocellular tumors [8–12].

Univariate and multivariate statistical methods can be applied to analyze NMR and MS peaks data or even on the metabolites identified from the data of these techniques and respective concentrations. Table 1 shows a selection of the most relevant studies in cancer metabolomics using NMR and MS. The data analysis strategies will be presented in the following sections.

2   Univariate Analysis

Univariate analysis studies a data variable at a time, crossing its values with those of metadata variables, being easy to perform and interpret, using methods such as t-tests (TT), one-way and multifactor analysis of variance (ANOVA), MannWhitney (MW), Kruskal-Wallis (KW) and Kolmogorov-Smirnov (KS) tests, fold change (FC), regression and correlation analysis (CA). These can provide sets of (ranked) variables, candidates for a better discrimination of a clinical variable. Thus, these techniques are quite useful for biomarker prediction, as well as a first step in classification or regression with machine learning.

Specifically in metabolomic cancer studies, univariate analysis has been performed in many studies as is clear from the previous table. One example is the use of one-way ANOVA and Tukey's Honest Significant Difference (HSD) test in studying NMR data from breast cancer cells [15]. Also, in a chemotherapy breast cancer study [33], the authors performed paired/unpaired t-tests over MS data, as well as two-way ANOVA to study the interaction of two variables.

3   Unsupervised Multivariate Analysis

This type of analysis summarizes data and thus detects patterns that can be related to biological or experimental variables.

Principal Component Analysis (PCA) is the most frequently used unsupervised learning method for data analysis, normally used in metabolomics to discover patterns in the data which may reveal how samples group based on their metabolic profiles. It is a dimensionality reduction technique, which produces new variables through linear combinations of the original variables [35], to explain as much of the variance in the original data set as possible.

In recent cancer studies using NMR, PCA has been applied, for instance, to discriminate between four groups of MCF7 breast cancer cell lines with or without tamoxifen resistance and/or CK-α downregulation [18], and to separate gastric cancer samples from control samples [3]. Regarding MS approaches, there are also some studies using PCA, for the detection of biomarkers related to prostate cancer, by combining it with supervised methods [36] or to access the different metabolic profiles of ovarian cancer stem cells and cancer cells [11].

On the other hand, Hierarchical clustering (HC) separates observations into groups and establishes a hierarchical ordering of the data points by taking into consideration a measure of dissimilarity between observations. In [15], HC was performed on metabolite concentration data derived from NMR experiments of different breast cancer cell lines to assess the effect of radiation therapy or poly ADP-ribose polymerase inhibition. In another study, the authors [13] used HC to evaluate the separation between advanced colorectal cancer samples and controls, based on data from NMR of fecal extracts. They did, however, conclude that PCA performed better at this task than HC. In [11], following a MS approach, HC demonstrated a clear separation between cell types, based on the intracellular profile of ovarian cancer stem cells and ovarian cancer cells, while in another MS cancer study, HC allowed the estimation of clinical metabolic biomarkers from plasma for diagnosis of esophageal squamous-cell carcinoma [37].

K-means is another clustering approach. It partitions observations into a pre-defined number k of groups. The algorithm is initialized considering k observations to be the initial clusters, and samples are assigned to the cluster with the nearest mean, recalculating the clusters after every assignment [38]. As an example of its application over MS data, in [39] the authors used it to identify metabolite signatures of malignant glioma from human cerebrospinal fluid.

4   Supervised Multivariate Analysis

On the other hand, supervised multivariate analysis creates models capable of predicting an output from a certain data input, based on data with known output.

Partial least squares (PLS) regression, partial least squares discriminant analysis (PLS-DA), and orthogonal partial least squares discriminant analysis (OPLS-DA) are the most popular supervised learning methods used in metabolomics studies. PLS [40] models the relationship between a matrix of predictor variables and one or more output variables by finding a set of new variables that maximize the explained covariance. PLS-DA is an adaptation of the partial least squares algorithm for classification, and is used to analyze group separation [41].

In recent metabolomics NMR and MS studies, PLS-DA has been used, for instance, to identify a urinary metabolite signature for renal cell carcinoma [19]. In [33], PLS-DA, using MS data, revealed a trend to separate premenopausal and postmenopausal samples, suggesting that altered serum levels of oleic acid in breast cancer patients are associated with their response to chemotherapy. OPLS-DA is a variant of PLS-DA in which non-correlated variation is removed to facilitate model interpretability [42]. It has been applied, for example, to discriminate between pancreatic adenocarcinoma and healthy tissue [4], and to differentiate between basal cell carcinoma and normal skin samples [21].

To build predictors in cancer metabolomics studies, random forests (RF) represent another model that can be used for classification or regression. RFs are ensembles of decision trees, which are made up of decision rules that are inferred from input data [43]. In an experiment, RF was used to determine if NMR data could distinguish between groups of cancer patients (with cachexia, pre-cachexia or weight stable) and healthy controls [26]. This RF was used as a feature selection step, evaluating the importance of each metabolite and subsequently selecting the fifteen most predictive metabolites. In another study using both NMR spectroscopy and MS [44], experimental data was used to train a RF that could distinguish between hepatocellular carcinoma, liver cirrhosis and control serum samples. The RF was valuable in selecting the most important metabolites that could accurately discriminate the groups and could be considered potential biomarkers. In another study, [9], RF models used MS data to train a set of lung cancer and control cases. The model revealed that three of the most highly well-known nicotine metabolites (cotinine, nicotine-N-oxide, and trans-3-hydroxycotinine) were the most important ones for the model to distinguish between both cases.

A Support Vector Machine (SVM) [45] is a machine learning method that maps input features to a new, linear feature space using a kernel function. Regarding NMR studies, [46] used a SVM with a radial basis function kernel to classify cell extracts from normal and hepatocellular carcinoma cell lines as well as the respective culture media. In [14], two supervised methods were combined - PLS was applied as a dimensionality reduction method and the resulting scores were used to train a SVM model to distinguish between patients with metastatic colorectal cancer and healthy individuals. In the same study, a PLS-SVM approach was also used to predict overall survival for the patients with metastatic colorectal cancer. In [47], SVM models were applied on MS data collected for sixteen diagnostic metabolites from lipid and fatty acid metabolism, allowing the identification of early-stage ovarian cancer patients.

5   Case Studies

Specmine [48] is an R package, developed in our group, for metabolomics data analysis that allows users to perform the analyses described in the previous sections, and many others. To demonstrate its usefulness in cancer metabolomics studies based on NMR and MS techniques, two studies were reproduced using the *specmine* package. The fully detailed reports can be accessed in the URL http://darwin.di.uminho.pt/PACBB2018/metabolomics.

The first study [49] analyzed the possible association of metabolism with the altered expression of the inositol 1,4,5 trisphosphate (IP3R) receptor in breast cancer, as this receptor is known to regulate metabolism and cellular bioenergetics and is upregulated in a number of cancers, by using the 1H CPMG NMR technique. Data for this analysis was obtained from the Metabolights website [50], under the study MTBLS152. The analysis performed included PCA and PLS-DA. Although there were some differences in terms of results, possibly due to the use of a dataset that is slightly different to the original file used by the authors, the specmine results confirm that PCA and PLS-DA were able to discriminate between samples with high and/or low expression of the gene that encodes inositol 1,4,5-trisphosphate receptor type 3 and healthy control samples.

The second study [11] analyzed the differences between ovarian cancer cells (OCCs) and cancer stem cells (OCSCs) as regards the intracellular and extracellular metabolomic profiles, by using the GC-MS technique. Data for this analysis was also obtained from Metabolights, under the study MTBLS152. The analysis performed included PCA and t-tests. Overall, the obtained results were very similar to the ones present in the article. Some of the differences may be due to the study authors not fully explaining how the analysis was conducted, especially regarding how they handled the fact that, in some cases, the same metabolite had different concentration levels for each sample.

6   Conclusions

Although the typical procedure in metabolomics data analysis usually involves PCA and PLS-DA/OPLS-DA analyses, most studies use a variety of data analysis methods that confirm and complement one another. Some recent cancer metabolomics studies have explored other machine learning techniques to build predictors based on NMR and/or MS data. These alternative predictors may be useful to build more robust classifiers and to extract biologically meaningful information from metabolomics data, such as identifying potential metabolic biomarkers. In the future, it would be interesting to see how these and other alternatives perform when compared to established methods.

Furthermore, with the reproduction of two studies using the *specmine* package, it is noticeable that this R package can be very useful in metabolomics data analysis, not only in univariate analysis, but also in multivariate analysis, such as machine learning and PCA.

Acknowledgments. This work is co-funded by the North Portugal Regional Operational Programme, under the "Portugal 2020", through the European Regional Development Fund (ERDF), within project SISBI- Ref^{a}NORTE-01-0247-FEDER-003381.

This study was also supported by the Portuguese Foundation for Science and Technology (FCT) under the scope of the strategic funding of UID/BIO/04469/2013 unit and COMPETE 2020 (POCI-01-0145-FEDER-006684) and BioTecNorte operation (NORTE-01-0145-FEDER-000004) funded by European Regional Development Fund under the scope of Norte2020 - Programa Operacional Regional do Norte.

References

1. Warburg, O.: On the origin of cancer cells. Science **123**(3191), 309–314 (1956)
2. Pavlova, N.N., Thompson, C.B.: The emerging hallmarks of cancer metabolism. Cell Metab. **23**(1), 27–47 (2016)
3. Wang, H., et al.: Tissue metabolic profiling of human gastric cancer assessed by 1H NMR. BMC Cancer **16**(1), 371 (2016)
4. Battini, S., et al.: Metabolomics approaches in pancreatic adenocarcinoma: tumor metabolism profiling predicts clinical outcome of patients. BMC Med. **15**(1), 56 (2017)
5. Hart, C.D., et al.: Serum metabolomic profiles identify ER-positive early breast cancer patients at increased risk of disease recurrence in a multicenter population. Clin. Cancer Res. **23**(6), 1422–1431 (2017)
6. Belkaid, A., et al.: Metabolic effect of estrogen receptor agonists on breast cancer cells in the presence or absence of carbonic anhydrase inhibitors. Metabolites **6**(2), 16 (2016)
7. Hao, D., et al.: Temporal characterization of serum metabolite signatures in lung cancer patients undergoing treatment. Metabolomics **12**(3), 58 (2016)
8. Fahrmann, J.F., et al.: Serum phosphatidylethanolamine levels distinguish benign from malignant solitary pulmonary nodules and represent a potential diagnostic biomarker for lung cancer. Cancer Biomarkers **16**(4), 609–617 (2016)
9. Mathé, E.A., et al.: Noninvasive urinary metabolomic profiling identifies diagnostic and prognostic markers in lung cancer. Cancer Res. **74**(12), 3259–3270 (2014)
10. Zhu, J., et al.: Colorectal cancer detection using targeted serum metabolic profiling. J. Proteome Res. **13**(9), 4120–4130 (2014)
11. Vermeersch, K., et al.: OVCAR-3 spheroid-derived cells display distinct metabolic profiles. PLoS One **10**(2), e0118262 (2015)
12. Ranjbar, M., et al.: GC-MS based plasma metabolomics for identification of candidate biomarkers for hepatocellular carcinoma in Egyptian cohort. PLoS One **10**(6), e0127299 (2015)
13. Amiot, A., et al.: 1 H NMR spectroscopy of fecal extracts enables detection of advanced Colorectal Neoplasia. J. Proteome Res. **14**(9), 3871–3881 (2015)

14. Bertini, I., et al.: Metabolomic NMR fingerprinting to identify and predict survival of patients with metastatic colorectal cancer. Cancer Res. **72**(1), 356–364 (2012)

15. Bhute, V.J., et al.: The poly (ADP-Ribose) polymerase inhibitor veliparib and radiation cause significant cell line dependent metabolic changes in breast cancer cells. Sci. Rep. **6**(1), 36061 (2016)

16. Chan, A.W., et al.: 1H-NMR urinary metabolomic profiling for diagnosis of gastric cancer. Br. J. Cancer **114**(1), 59–62 (2016)

17. Fages, A., et al.: Metabolomic profiles of hepatocellular carcinoma in a European prospective cohort. BMC Med. **13**(1), 242 (2015)

18. Kim, H.S., et al.: Investigation of discriminant metabolites in tamoxifen-resistant and choline kinase-alpha-downregulated breast cancer cells using 1H-nuclear magnetic resonance spectroscopy. PLoS One **12**(6), e0179773 (2017)

19. Monteiro, M.S., et al.: Nuclear magnetic resonance metabolomics reveals an excretory metabolic signature of renal cell carcinoma. Sci. Rep. **6**(1), 37275 (2016)

20. Morin, P.J., et al.: NMR metabolomics analysis of the effects of 5-lipoxygenase inhibitors on metabolism in glioblastomas. J. Proteome Res. **12**(5), 2165–2176 (2013)

21. Mun, J., et al.: Discrimination of basal cell carcinoma from normal skin tissue using high-resolution magic angle spinning 1H NMR spectroscopy. PLoS One **11**(3), e0150328 (2016)

22. Roberts, M.J., et al.: Seminal plasma enables selection and monitoring of active surveillance candidates using nuclear magnetic resonance-based metabolomics: a preliminary investigation. Prostate Int. **5**(4), 149–157 (2017)

23. Shao, W., et al.: Malignancy-associated metabolic profiling of human glioma cell lines using 1H NMR spectroscopy. Mol. Cancer **13**(1), 197 (2014)

24. Tsai, I., et al.: Metabolomic dynamic analysis of hypoxia in MDA-MB-231 and the comparison with inferred metabolites from transcriptomics data. Cancers **5**(2), 491–510 (2013)

25. Uifălean, A., et al.: The impact of Soy Iso avones on MCF-7 and MDA-MB-231 breast cancer cells using a global metabolomic approach. Int. J. Mol. Sci. **17**(9), 1443 (2016)

26. Yang, Q.J., et al.: Serum and urine metabolomics study reveals a distinct diagnostic model for cancer cachexia. J. Cachexia Sarcopenia Muscle **9**(1), 1–15 (2017)

27. Miyamoto, S., et al.: Systemic metabolomic changes in blood samples of lung cancer patients identified by gas chromatography time-of-flight mass spectrometry. Metabolites **5**(2), 192–210 (2015)

28. Batova, A., et al.: Englerin A induces an acute inflammatory response and reveals lipid metabolism and ER stress as targetable vulnerabilities in renal cell carcinoma. PLoS One **12**(3), e0172632 (2017)

29. Xiao, J.F., et al.: LC-MS based serum metabolomics for identification of hepatocellular carcinoma biomarkers in Egyptian cohort. J. Proteome Res. **11**(12), 5914–5923 (2012)

30. Dhakshinamoorthy, S., et al.: Metabolomics identifies the intersection of phosphoethanolamine with menaquinone-triggered apoptosis in an in vitro model of leukemia. Mol. BioSyst. **11**(9), 2406–2416 (2015)

31. Mackay, E.: Fatty acid synthesis in colorectal cancer: characterization of lipid metabolism in serum, tumour, and normal host tissues. Ph.D. thesis, University of Calgary (2015)

32. Vermeersch, K.A., et al.: Distinct metabolic responses of an ovarian cancer stem cell line. BMC Syst. Biol. **8**(1), 134 (2014)

33. Hilvo, M., et al.: Monounsaturated fatty acids in serum triacylglycerols are associated with response to neoadjuvant chemotherapy in breast cancer patients. Int. J. Cancer **134**(7), 1725–1733 (2014)
34. Ressom, H.W., et al.: Utilization of metabolomics to identify serum biomarkers for hepatocellular carcinoma in patients with liver cirrhosis. Analytica Chimica Acta **743**, 90–100 (2012)
35. Abdi, H., Williams, L.J.: Principal component analysis. Wiley Interdisc. Rev. Comput. Stat. **2**(4), 433–459 (2010)
36. Zhang, T., et al.: Application of holistic liquid chromatography-high resolution mass spectrometry based urinary metabolomics for prostate cancer detection and biomarker discovery. PLoS One **8**(6), e65880 (2013)
37. Liu, R., et al.: Identification of plasma metabolomic profiling for diagnosis of esophageal squamous-cell carcinoma using an UPLC/TOF/MS platform. Int. J. Mol. Sci. **14**(5), 8899–8911 (2013)
38. Macqueen, J.: Some methods for classification and analysis of multivariate observations. In: Proceedings of the Fifth Berkeley Symposium on Mathematical Statistics and Probability, vol. 1(233), pp. 281–297 (1967)
39. Locasale, J.W., et al.: Metabolomics of human cerebrospinal fluid identifies signatures of malignant glioma. Mol. Cellular Proteomics **11**(6), M111.014688 (2012)
40. Wold, H.: Estimation of principal components and related models by iterative least squares. In: Multivariate Analysis, pp. 1391–1420 (1966)
41. Barker, M., Rayens, W.: Partial least squares for discrimination. J. Chemom. **17**(3), 166–173 (2003)
42. Trygg, J., Wold, S.: Orthogonal projections to latent structures (OPLS). J. Chemom. **16**(3), 119–128 (2002)
43. Breiman, L.: Random forests. Mach. Learn. **45**(1), 5–32 (2001)
44. Liu, Y., et al.: NMR and LC/MS-based global metabolomics to identify serum biomarkers differentiating hepatocellular carcinoma from liver cirrhosis. Int. J. Cancer **135**(3), 658–668 (2014)
45. Cortes, C., Vapnik, V.: Support-vector networks. Mach. Learn. **20**(3), 273–297 (1995)
46. Chen, Y., et al.: Metabolic profiling of normal hepatocyte and hepatocellular carcinoma cells via 1 H nuclear magnetic resonance spectroscopy. Cell Biol. Int. **9999**, 1–10 (2017)
47. Gaul, D.A., et al.: Highly-accurate metabolomic detection of early-stage ovarian cancer. Sci. Rep. **5**, 16351 (2015)
48. Costa, C., et al.: An R package for the integrated analysis of metabolomics and spectral data. Comput. Meth. Prog. Biomed. **129**, 117–124 (2016)
49. Singh, A., et al.: 1H NMR metabolomics reveals association of high expression of inositol 1,4,5 trisphosphate receptor and metabolites in breast cancer patients. PLoS One **12**(1), 1–20 (2017)
50. Haug, K., et al.: MetaboLights - an open-access general-purpose repository for metabolomics studies and associated meta-data. Nucleic Acids Res. **41**(D1), D781–D786 (2013)

A Framework for the Automatic Combination and Evaluation of Gene Selection Methods

Bastien Grasnick, Cindy Perscheid$^{(\boxtimes)}$, and Matthias Uflacker

Hasso Plattner Institute, Potsdam, Germany
cindy.perscheid@hpi.de

Abstract. High-throughput RNA-Sequencing technologies produce large gene expression datasets whose analysis leads to a better understanding and treatment of diseases like cancer. The data's high dimensionality poses challenges to its computational analysis, which is addressed by applying gene selection. Traditional gene selection methods are based on the data only. In turn, integrative approaches include curated biological information from external knowledge bases in the gene selection process, which improves result accuracy and computational complexity.

This paper presents a framework for comparing knowledge based and computational gene selection. Moreover, a novel integrative method of the automatic combination of both approaches is presented. Results on a cancer dataset show that simple computational methods enriched by external knowledge can compete with complex computational techniques.

Keywords: Gene expression data analysis · Gene selection
Pattern recognition

1 Introduction

Recent advances in high-throughput sequencing technologies allow a more in-depth analysis of the human genome. RNA-Sequencing, in particular, measures the gene activity, i.e. gene expression levels, from vast amounts of genes from specific tissues [17]. A major part of research using this data is investigating the inner workings and mechanisms of diseases. The analysis of the interplay between genes and their connections to diseases leads to a deeper understanding of the matter, helps to improve diagnosis and results in more effective treatment [28]. The high dimensionality of gene expression data containing tens of thousands of genes as features coupled with a low number of samples poses challenges to its processing. Data analysis requires dimensionality reduction for a feasible application of computational methods like machine learning. Reducing the feature space is reasonable because the data is noisy and redundant [2]. Feature selection is a technique that addresses these problems. In the context of analyzing gene expression data, it is referred to as gene selection for the remainder of this

© Springer Nature Switzerland AG 2019
F. Fdez-Riverola et al. (Eds.): PACBB 2018, AISC 803, pp. 166–174, 2019.
https://doi.org/10.1007/978-3-319-98702-6_20

work. Features, i.e. genes, are ranked according to a specific measure, which allows to select the most valuable genes.

Computational approaches for gene selection identify the most valuable genes according to the patterns they find in the data. These patterns should typically reflect biological processes, which are highly complex. Identifying these processes correctly from the data alone can be a hard task: For example, two genes have similar expression patterns in the data, but participate in different biological processes. Computational approaches for gene selection would see a correlation of these two genes, thus interpreting they share the same process and select only one of them. Therefore, it is crucial to include the biological context early into the analysis. Literature also suggests that the integration of biological knowledge has advantages such as the supplement of data analysis, increased performance and the discovery of connections between data and external knowledge [3]. At the same time, machine-readable, curated repositories and meta-repositories of biological knowledge in the form of gene-disease-associations, pathway information and ontologies already exist as external knowledge bases [6,15,23]. We can use this information to enrich computational gene selection approaches.

The aim of this paper is to investigate how useful it is to integrate external knowledge bases in gene selection. We compare proven computational approaches for gene selection and combine them in an efficient manner with external knowledge bases. Our main contribution is two-fold: On the one hand, we present a framework for the automated and flexible comparison of various gene selection methods. On the other hand, we introduce a novel approach to flexibly combine computational gene selection methods with external knowledge.

The rest of the paper is structured as follows. Sect. 2 presents related work. Section 3 presents our framework. Section 4 describes the experiments, whose results are discussed in Sect. 5. Section 6 concludes our findings.

2   Related Work

Literature classifies gene selection approaches into six categories: Filter, wrapper, embedded, hybrid, ensemble, and integrative [2,3]. Filters are simple statistical methods and often applied for feasibility reasons, while the others apply machine learning methods – wrapper & embedded – or combine multiple gene selection methods – hybrid & ensemble – for more accurate results. Integrative gene selection incorporates domain knowledge from external knowledge bases, which allows better biological interpretation of the chosen genes, further motivating the integration of external knowledge [3,11]. Table 1 provides a classification overview of recent gene selection approaches. In the following, we concentrate on integrative gene selection approaches.

Qi and Tang integrate Gene Ontology (GO) into gene selection [24]. Genes are ranked by discriminative scores and annotated with GO terms. These terms are then ranked by the average discriminative scores of their associated genes. The final gene set is chosen by iteratively selecting the highest ranked genes from the highest ranked GO terms. Quanz et al. use KEGG pathways as features instead of genes and apply logistic regression to transform a pathway into

Table 1. Overview on gene selection approaches focused on the bioinformatics domain

Type	Functionality	Characteristics	Examples
Filter	Uses only intrinsic data characteristics	+ independent of classifier + low complexity + good generalization	mRMR [9] ReliefF [16] Information Gain (IG) [7]
Wrapper	Uses learning algorithm to evaluate genes	+ detects gene dependencies high complexity/overfitting	Genetic Algorithms (GA) [21] Successive gene selection (SFS) [27]
Embedded	Gene selection embedded in learning algorithm	+ detects gene dependencies / interacts with classifier	SVM-RFE [13] Random Forest [8]
Hybrid	Apply multiple approaches sequentially	/ intermediate complexity − risk of slight overfitting	SVM-RFE + mRMR Filter [20] MFMW gene selection [18]
Ensemble	Uses aggregate of group of gene sets	+ less prone to overfitting − computationally expensive − difficult to interpret	Ensemble Gene Selection By Grouping (EGSG) [19] MCF-RFE [30]
Integrative	Integrate external knowledge for gene selection	+ interpretable for experts − domain-specific − dependent on prior knowledge	KOGS [31] SoFoCles [22] and [1, 11, 24–26]

a feature value [15, 25]. Knowledge-Oriented Gene Selection (KOGS) integrates external knowledge sources and a combination of gene rankings [31]. They map different external sources to internal knowledge as sample geometric pattern, gene connection and gene function. Genes are selected via geometric consistency checking, relevance propagating or functional relevance voting. The combination of rankings is achieved by a probabilistic model that learns combination coefficients. SoFoCles annotates genes with GO knowledge to find semantically similar genes [22]. It ranks genes by a discriminative score, annotates all genes with GO terms, and computes the semantic similarity of the genes based on their GO terms. The final gene set consists of top ranked genes and those that are semantically most similar to them. Fang et al. combine Information Gain (IG) with knowledge from KEGG and GO using association analysis [11]. Raghu et al. find maximally relevant and diverse gene sets with Preferential Diversity (PrefDiv) using an importance score that combines prior knowledge and data inherent information [26]. Acharya et al. replace genes by their GO terms as features [1]. Genes are mapped to multiple related GO terms and a gene-GO term matrix is constructed. On it, Partitioning Around Medoids (PAM) clustering is applied for feature selection using Euclidean distance, city block or cosine distance.

The presented approaches are highly specific to a single knowledge base, e.g. KEGG or GO. Most of them again apply machine learning techniques, which increases computational complexity. In contrast, our framework flexibly combines low complexity computational gene selection approaches with different knowledge bases and evaluates the results with multiple standard metrics.

3   Proposed Framework

To investigate the usefulness of the integration of biological knowledge in gene selection for gene expression data, we developed a framework for the comparison of computational, knowledge based and combined gene selection methods.

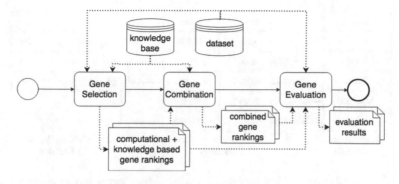

Fig. 1. Three step integrative gene selection, combination and evaluation framework

The framework takes as input a gene expression dataset with some meta data, e.g. associated disease, and an external knowledge base. It outputs ranked lists of genes from computational, knowledge based and combined methods with their respective evaluation results, e.g. clustering accuracy. Figure 1 depicts the distinct steps in the framework: gene selection, gene combination and gene evaluation.

Gene Selection. This step receives a gene expression dataset and information from external knowledge bases as input. It further splits up into the application of computational gene selection techniques and our knowledge based gene selection approach. It outputs computational and knowledge based gene rankings. Firstly, the computational gene selection part produces gene rankings based on statistical methods or machine learning. The user chooses the respective gene selection method. Secondly, the knowledge based gene selection approach creates a gene ranking based on information from external knowledge bases. The user as well chooses the respective knowledge base. In our current prototype, we have integrated KEGG and DisGeNET as external knowledge bases. DisGeNET integrates data from various scientific biological sources containing information on >560 K gene-disease-associations (GDAs) [23]. Our framework produces the gene ranking by a) extracting a list of diseases from the dataset, b) fetching a list of genes for each disease, e.g. sorted by their GDA score, c) filtering these lists by a user-defined top k or threshold value, and d) combining all lists into one. We combine gene lists in an interleaving manner, this way preventing diseases with a long gene list or highly-scored genes to dominate the final gene list and representing each disease equally.

Gene Combination. The gene combination step uses the produced gene rankings as well as prior knowledge as inputs and returns combined gene rankings. As shown in Fig. 2, our approach filters the computational gene rankings by keeping only genes that biological knowledge identified as connected to the diseases apparent in the dataset. We use DisGeNET and KEGG as knowledge bases in this step. For DisGeNET, we use all genes contained in the DisGeNET gene

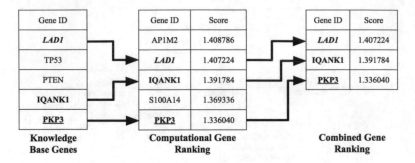

Fig. 2. Combined gene ranking by using knowledge base extracted genes as a filter for the computational gene ranking

ranking as an unranked set of genes. For KEGG, we use all genes that are contained in pathways connected to the diseases extracted from the dataset.

Gene Evaluation. This step receives the gene expression dataset and all previously produced gene rankings as input and returns the respective evaluation results. The framework retains the top n genes from each ranking, with $n = 2...100$. It then applies multiple classifiers for training and evaluation on the reduced gene expression data and averages their classification performance metrics.

Technical Implementation. Our framework is mainly implemented in Python and integrates WEKA and R. It has dedicated Python modules for accessing the APIs of the external knowledge bases, at the time being KEGG and DisGeNET. The framework owns modules for accessing WEKA and R, which provides users the flexibility to use any computational gene selection, classifier and evaluation metric that is offered by WEKA or R [14]. If users want to integrate a new computational gene selection method that is not present in R or WEKA, they have to integrate it there or implement a new Python module in our framework. The same goes for the integration of new knowledge bases. The framework is available on https://github.com/CPerscheid/GSEF.git.

4   Experiments

For the experiments, we selected cancer data from The Cancer Genome Atlas (TCGA) for eight cancer types because of their good coverage in DisGeNET (GBM, HNSC, KIRC, LAML, LUAD, SARC, THCA, UCEC) [12,29]. We filtered out all genes and samples that had more than 30% missing values. Afterwards, the data was normalized by library size and log-transformed using the VST procedure [10]. After preprocessing, the dataset consisted of 3,189 samples and expression values for 55,572 genes. First, we ran computational and DisGeNET based gene selection on this dataset. We selected our computational gene selection approaches according to recommendations from

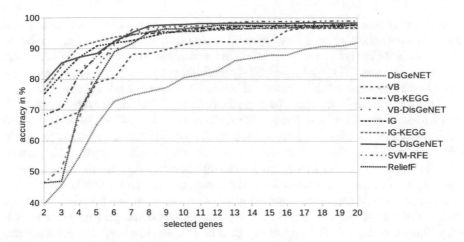

Fig. 3. Combination gene ranking by using knowledge base extracted genes as a filter for the computational gene ranking

Bolón-Canedo et al. [4,5]. They conclude that IG, ReliefF and SVM-RFE are the best approaches in their respective categories of univariate filter, multivariate filter and embedded. We did not use any wrapper approach because of their high complexity and no clear advantages over filter or embedded methods. Additionally, we used a variance-based gene selection (VB) because it is fast and well accepted in differential gene expression workflows. Next, we produced combinations of the gene rankings from VB and IG with the unranked gene sets from the knowledge bases DisGeNET and KEGG. We chose VB and IG for the combination as they are simple methods that quickly rank genes. All gene rankings were evaluated on the dataset using 10-fold cross validation with support vector machines (SVM), logistic regression, naive Bayes and k-nearest neighbors algorithm (KNN, with $k = 3; 5$) by averaging their classification accuracy.

Figure 3 compares DisGeNET based gene selection, computational methods and combined approaches. All computational methods outperform DisGeNET with more than four genes used. From the computational methods, IG performs best up until six genes selected while after that SVM-RFE achieves the best performance. Moreover, all combination approaches perform better than the DisGeNET method alone. The VB combinations perform similarly and both improve over using only VB for gene selection. The same is true for IG combinations, which perform best up until 12 genes. For more than 12 genes, SVM-RFE performs best. At 17+ genes, all approaches apart from DisGeNET plateau and perform similarly with accuracies between 96% and 98%.

5   Discussion

As our results show, purely knowledge based gene selection gets outperformed by computational methods. However, enriching simple methods like VB and IG with prior knowledge improves performance. Every combination is able to compete with complex methods like SVM-RFE. Another advantage of the combined approach is its low complexity and thus low execution time. This makes it useful for large datasets. The biological interpretability of the chosen genes by using prior knowledge and simple statistical methods is another advantage.

There are some limitations to the proposed combined approach though. We require knowledge of which diseases are contained in the dataset to map them to the knowledge bases for the automatic selection of biologically relevant genes. In addition, sufficient information about those diseases needs to exist in the knowledge bases. However, as biological knowledge bases gather more data through research on diseases like cancer and Alzheimer, our approach will become more viable and useful in an increasing number of scenarios in the future. To further verify our results, the combined gene selection method integrating prior knowledge should be tested with other datasets and other diseases. A more in-depth study with more computational methods and integrating more prior knowledge could show even better results for combinations or reveal cases where the integration of biological knowledge is especially useful. Our framework enables and encourages this as it can be easily extended by further gene selection methods and external knowledge sources.

6   Conclusion

This paper presented a framework for comparing both computational and knowledge based gene selection methods for gene expression data. Moreover, a novel integrative method of combining both approaches was presented. The results show that while the knowledge based approach performs worse than computational techniques, a combination of both is promising. The combination of simple computational methods with external knowledge leads to performance improvements and can compete with more complex methods that might be unfeasible given a dataset that is very large.

References

1. Acharya, S., Saha, S., Nikhil, N.: Unsupervised gene selection using biological knowledge: application in sample clustering. BMC Bioinform. **18**(1), 513 (2017)
2. Ang, J.C., et al.: Supervised, unsupervised, and semi-supervised feature selection: a review on gene selection. IEEE/ACM Trans. Comput. Biol. Bioinform. **13**(5), 971–989 (2016)
3. Bellazzi, R., Zupan, B.: Towards knowledge-based gene expression data mining. J. Biomed. Inform. **40**(6), 787–802 (2007)
4. Bolón-Canedo, V., et al.: A review of microarray datasets and applied feature selection methods. Inf. Sci. **282**, 111–135 (2014)
5. Bolón-Canedo, V., Sánchez-Maroño, N., Alonso-Betanzos, A.: A review of feature selection methods on synthetic data. Knowl. Inf. Syst. **34**(3), 483–519 (2013)
6. Consortium, G.O., et al.: Expansion of the Gene Ontology knowledgebase and resources. Nucleic Acids Res. **45**(D1), D331–D338 (2017)

7. Dash, M., Liu, H.: Feature selection for classification. Intell. Data Anal. **1**(3), 131–156 (1997)
8. Díaz-Uriarte, R., De Andres, S.A.: Gene selection and classification of microarray data using random forest. BMC Bioinform. **7**(1), 3 (2006)
9. Ding, C., Peng, H.: Minimum redundancy feature selection from microarray gene expression data. J. Bioinform. Comput. Biol. **3**(02), 185–205 (2005)
10. Durbin, B.P., et al.: A variance-stabilizing transformation for gene-expression microarray data. Bioinformatics **18**(suppl. 1), S105–S110 (2002)
11. Fang, O.H., Mustapha, N., Sulaiman, M.N.: An integrative gene selection with association analysis for microarray data classification. Intell. Data Anal. **18**(4), 739–758 (2014)
12. Grossman, R.L., et al.: Toward a shared vision for cancer genomic data. N. Engl. J. Med. **375**(12), 1109–1112 (2016)
13. Guyon, I., et al.: Gene selection for cancer classification using support vector machines. Mach. Learn. **46**(1–3), 389–422 (2002)
14. Hall, M., et al.: The WEKA data mining software: an update. SIGKDD Explor. **11**(1), 10–18 (2009)
15. Kanehisa, M., Goto, S.: KEGG: Kyoto encyclopedia of genes and genomes. Nucleic Acids Res. **28**(1), 27–30 (2000)
16. Kononenko, I.: Estimating attributes: analysis and extensions of RELIEF. In: European Conference on Machine Learning, pp. 171–182. Springer (1994)
17. Kukurba, K.R., Montgomery, S.B.: RNA sequencing and analysis. Cold Spring Harb. Protoc. **2015**(11), 951–69 (2015)
18. Leung, Y., Hung, Y.: A multiple-filter-multiple-wrapper approach to gene selection and microarray data classification. IEEE/ACM Trans. Comput. Biol. Bioinform. **7**(1), 108–117 (2010)
19. Liu, H., Liu, L., Zhang, H.: Ensemble gene selection by grouping for microarray data classification. J. Biomed. Inform. **43**(1), 81–87 (2010)
20. Mundra, P.A., Rajapakse, J.C.: SVM-RFE with MRMR filter for gene selection. IEEE Trans. Nanobiosci. **9**(1), 31–37 (2010)
21. Ooi, C., Tan, P.: Genetic algorithms applied to multi-class prediction for the analysis of gene expression data. Bioinformatics **19**(1), 37–44 (2003)
22. Papachristoudis, G., Diplaris, S., Mitkas, P.A.: SoFoCles: feature filtering for microarray classification based on gene ontology. J. Biomed. Inform. **43**(1), 1–14 (2010)
23. Piñero, J., et al.: DisGeNET: a discovery platform for the dynamical exploration of human diseases and their genes. Database **2015** (2015)
24. Qi, J., Tang, J.: Integrating gene ontology into discriminative powers of genes for feature selection in microarray data. In: Proceedings of APGV. ACM (2007)
25. Quanz, B., Park, M., Huan, J.: Biological pathways as features for microarray data classification. In: International Workshop on Data and Text Mining in Biomedical Informatics, pp. 5–12. ACM (2008)
26. Raghu, V.K., et al.: Integrated theory-and data-driven feature selection in gene expression data analysis. In: Proceedings of International Conference on Data Engineering, pp. 1525–1532. IEEE (2017)
27. Sharma, A., Imoto, S., Miyano, S.: A top-r feature selection algorithm for microarray gene expression data. IEEE/ACM Trans. Comput. Biol. Bioinform. **9**(3), 754–764 (2012)
28. Soh, D., et al.: Enabling more sophisticated gene expression analysis for understanding diseases and optimizing treatments. SIGKDD Explor. **9**(1), 3–13 (2007)

29. Weinstein, J.N., et al.: The cancer genome atlas pan-cancer analysis project. Nat. Genet. **45**(10), 1113–1120 (2013)
30. Yang, F., Mao, K.: Robust feature selection for microarray data based on multicriterion fusion. IEEE/ACM Trans. Comput. Biol. Bioinform. **8**(4), 1080–1092 (2011)
31. Zhao, Z., et al.: An integrative approach to identifying biologically relevant genes. In: Proceedings of SIAM International Conference Data Mining 2010, pp. 838–849. SIAM (2010)

Parallel Cellular Automaton Tumor Growth Model

Alberto G. Salguero[1]([✉]), Manuel I. Capel[2], and Antonio J. Tomeu[1]

[1] University of Cádiz, Cádiz, Spain
{alberto.salguero,antonio.tomeu}@uca.es
[2] University of Granada, Granada, Spain
manuelcapel@ugr.es

Abstract. "In silico" experimentation allows us to simulate the effect of different therapies by handling model parameters. Although the computational simulation of tumors is currently a well-known technique, it is however possible to contribute to its improvement by parallelizing simulations on computer systems of many and multi-cores. This work presents a proposal to parallelize a tumor growth simulation that is based on cellular automata by partitioning of the data domain and by dynamic load balancing. The initial results of this new approach show that it is possible to successfully accelerate the calculations of a known algorithm for tumor-growth.

Keywords: Cellular automaton · High performance computing
Mathematical oncology · Tumoral growth simulation
Parallel programming · Speedup

1 Introduction

Tumor growth from a transformed cancer-cell into a clinically apparent mass spans a range of spatial and temporal magnitudes. Cellular Automata (CA) can accurately describe the complexity of tumor development [6,20] and this is reproduced by computer simulation. The development of appropriate CA-based software tools will enable tumor prognosis without the need for patients to undergo annoying medical examinations or painful invasive tests.

In order to speed up these computer simulations, recent contributions [16] show advanced techniques for modelling tumor growth such as the use of efficient data structures for supporting deterministic cellular automata (DCA). Multi-paradigm and multiscale models of cancer dynamics [13] have been developed to predict tumor growth and therapeutics.

There have been some approaches based on CA optimization to further extend multiparadigm tumor growth models [2,3,12,14,15,17] and these mainly aim to improve computer simulation performance by guaranteeing efficient data memory program access [16], or by considering the dynamic evolution of the memory space (grids, trees...) that holds crucial data in simulations [18].

© Springer Nature Switzerland AG 2019
F. Fdez-Riverola et al. (Eds.): PACBB 2018, AISC 803, pp. 175–182, 2019.
https://doi.org/10.1007/978-3-319-98702-6_21

In our opinion, the different optimizations based on improving sequential data structure access are not decisive enough to achieve the high-performance computing power actually required by programs simulating cell behaviour. The possibility of using multicore and GPU parallelism as a promising multiplatform and framework to develop new programming techniques to speed-up the simulation computation time has only just started to be explored in recent years [5,8,11].

In order to be able to speed up in parallel programs, this paper presents a CA-based model for tumor growth simulation and identifies the synchronization instructions for implementing this model in a multicore processor in Java.

2   Tumor Growth Model

There are various mathematical definitions of cellular automata. We chose the one established in [1] as the most generally accepted in Computational Sciences, adapted to represent lattice-based biological models [4], and applied to simulate dynamic tumor growth in [16].

One tumor cell in the model is an individual entity that takes up one node of a finite 2D lattice ζ and which can carry out the following actions: migrate to another node on the grid, proliferate by mitosis, die or remain quiescent. A live cell can proliferate by generating a cell's daughter through mitosis whenever there is space available in its neighborhood (given by the Moore neighborhood).

According to Poleszczuk-Enderling's model [16] of reference, the most efficient way of processing tumoral cells consists in keeping a linear list of the tumor cells in the lattice that occupy sequential positions in memory. The list is entirely processed in each simulation step and this yields a new list of cells to be processed in the next one and so on and so forth.

The lattice is used here merely to keep the current state of the tumor. Only the changes in cells are actually written to it. Only two lists are in fact needed to implement the tumoral growth simulation procedure: one listing the cells still to be processed and the second one storing the tumor cells for the next simulation step. This scheme is similar to the one proposed by [10] and which is, to the best of our knowledge, the best sequential solution to the tumor growth simulation problem to date. However, our model uses multicore processors to improve the speed of the tumor growth simulation.

In order to speed up the tumor growth simulation model, two fundamental problems must be solved: firstly, to find a good strategy to maintain balanced cell distribution between threads; and secondly, to prevent access to the data structures (the lattice and lists of active cells) by each thread from blocking other threads for an unacceptable amount of time that could worsen program performance. It is important to note that insertions and deletions on the lattice cell lists must be performed under mutual exclusion to avoid loss of information.

By using different lists for each region it is possible to prevent blocking access to the *current-state* cell list. The *next-state* lists must always be accessed by using blocking access primitives, since cells can migrate between regions when program threads access such lists.

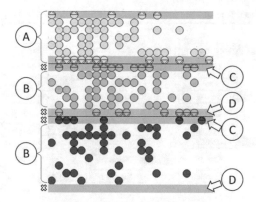

Fig. 1. Lattice division: (A) region assigned to a unique thread; (B) center subregion; (C) bottom part of the seam subregion; (D) top part of the seam subregion

The main objective of the model proposed in this work is to enable concurrent processing of the cells without the need for programming blocking access both to the cell lists and the lattice that holds the current state of the tumor mass. In order to achieve this purpose, the lattice is first divided vertically into as many regions as threads available. Each of these regions is further divided into three parts or subregions: top seam part (C), central part (B) and bottom seam part (D), as shown in Fig. 1. The central parts of the regions are therefore separated by what we call "seams", which are divided into two parts. This layout guarantees the existence of at least two types of subregions between subregions of the same type. This allows the model to concurrently process the cells in subregions of the same type without needing to block access since threads do not enter subregions which are being accessed by other threads. In the case of cell migration to another region, no other thread may access the target subregion at the same time. In order to make use of this property, the proposed model first processes the cells in the bottom seam subregions, then processes those in the central parts and finally those in the top seam subregions (see Fig. 2). All this processing is carried out concurrently and without blocking the threads, except for making them wait until the remaining threads have processed the same type of subregion.

For the model to work efficiently, it is important to perform a correct load balancing between the number of cells that each thread obtains for processing. Once the entire next state of the tumor has been calculated, the main thread is responsible for adjusting the regions, if appropriate.

Subregions must have a minimum height so that no single cell can migrate between non-adjacent subregions. In the proposed model, the minimum height value is δ, the length of the longest displacement of a cell in each simulation step.

2.1   Seams Adjustment During Simulation

As the tumor grows, it is necessary to adjust the seams to equally distribute cells among all the threads. The main thread is responsible for performing this task

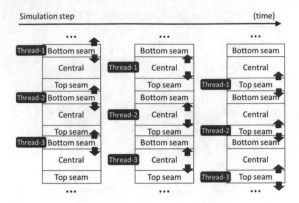

Fig. 2. A step in the parallel simulation

once the remaining threads have finished calculating the next state of the simulation. For reasons of efficiency, the main thread does not update the subregion to which each cell belongs. When the cells are again evaluated by the threads in the next step, the cells are assigned to the corresponding subregion. However, this may cause a cell to directly migrate from the top part of a seam (C) in one region to the central part of another region without passing through the bottom seam subregion, as shown in Fig. 3a. In substep 2, when the upper seam has been moved upward, the a cell still remains in the list of the top seam part (C). In substep 3, both threads simultaneously process the top seam parts (C), and this may cause incorrect concurrent access to the list of cells in the center area (B) or the need to use blocking access primitives. To avoid this situation, the seams are displaced in a three-tier process that is performed in three[1] consecutive and different steps of the simulation, as shown in Fig. 3b:

1. Increase the thickness of the seam in δ rows in the same direction of the displacement (step 2 in Fig. 3b).
2. Decrease and increase the size of the subregions that share the seam in δ rows: the part of the seam that was increased in the previous step is decreased and the other part is increased, maintaining the same seam thickness (step 4 in Fig. 3b).
3. Decrease the thickness of the seam to the default value in the region that has increased its size (step 5 in Fig. 3b).

   In this case, cell a can be safely added to list (B) in substep 6 because cell b cannot be processed in the same substep (they are still in different subregions). Since both cells have been assigned to the same thread in substep 3, they cannot be processed at the same instant. For this reason, there is no need to use blocking access primitives in substep 7, when both cells modify the center list (B).

---

[1] For the sake of simplicity, only substeps where changes have been made to the lists are shown in Fig. 3. Substeps 3 and 4 are actually part of the same overall step.

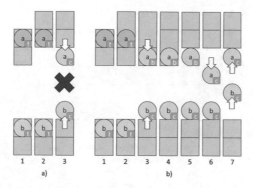

Fig. 3. Incorrect (a) and correct (b) parallel seam adjustment. The small letters in rectangles indicate the current list the cells belong to (t = top seam part, b = bottom seam part, c = center).

The seam adjusting process is automatically started when a significant difference of cells between two adjacent regions is detected. After some tests, it has been observed that the best results are obtained when this difference is greater than 5–15%. The exact value depends on the number of threads in the simulation.

It should also be noted that for reasons of efficiency it is not possible to adjust two consecutive seams. Since the main thread does not reassign migrated cells, it is not possible to determine the number of remaining cells in affected regions. Although this value could easily be calculated, it is important to note that the remaining threads are blocked until the main thread adjusts the seams, so it is crucial for seam adjustment to be performed as quickly as possible.

3   Model Implementations and Measurements

Some experiments have been conducted in order to verify the efficiency of the solution proposed in this work. In these experiments we measured the time taken by a Java application to follow the proposed model and we compared the results with the best known implementation to date [10]. A single cell has been placed at the center of the lattice in every case. The lattice in processed by the threads in parallel, according to model explained in previous section. The application developed follows the Poleszczuk-Enderling model for the simulation of tumoral growth. The model decides whether a cell dies, remains quiescent, or proliferates by mitosis (if there is enough space around the cell to do so) according to a set of probability distributions, which enables us to use a simulation based on a Monte Carlo stochastic submodel. We have employed the default parameters values provided by Poleszczuk-Enderling. More specifically: the cells may be divided ten times before proliferative capacity exhaustion; the probability of division is 1/24; the probability of symmetric division is 0.1; the probability of spontaneous death is 1/24; and the probability of migration is 10/24.

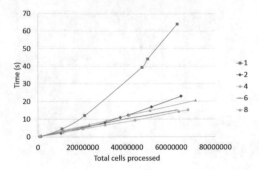

Fig. 4. Execution time for an Intel Core i7 implementation on 4 cores and 8 threads (hyperthreading activated) of the EP-parallel tumor growth simulation

Since the final result in the concurrent version depends on the order in which the cells are evaluated by the threads, it is never possible to obtain the exact same result. The shape of the resultant tumor mass is roughly the same for the same initial configuration and simulation parameters. However, since the border of the tumor is diffuse, the exact number of cells in the tumor mass may vary a lot. Therefore, the time spent on each simulation step cannot be compared among different simulations. Instead, the number of cells processed by time unit is used as reference. Five different configurations with $n = \{1, 2, 4, 6, 8\}$ threads have been used on the i7, while seven different configurations with $n = \{1, 2, 4, 6, 8, 10, 12\}$ threads have been used on the Xeon.

Figure 4 shows that a huge number of cells must be processed in each simulation. Around 300 simulation steps are necessary for the tumor to reach the cell-lattice border regions for the case of $n = 8$ threads, for example. The threads associated with these regions are idle until then. The load will only be equally

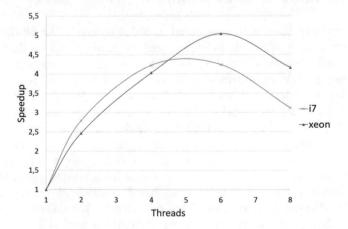

Fig. 5. Speed-up gain for Xeon E5-2670 and Core i7-6700 of EP-parallel tumor growth with respect to the sequential EP simulation

distributed among all the threads after 500–600 simulation steps. The speed-up gain for $\{2,4,6,8\}$-thread configurations is shown in Fig. 5. The same behavior is also observed when the experiment is performed in one of the nodes of the University of Cádiz. No noticeable improvement has been found for more than six threads.

4   Conclusion and Future Work

The application of parallelization techniques to cellular automata-based tumor growth simulations on multicore and many core or processor clusters is a research field which has not yet reached full maturity. By introducing a data parallel scheme with several computing threads, we have proposed a parallelization approach for a basic tumor growth simulation model. Different processing loads were deployed on selected multicore processor architectures for a Java parallel program and compared with the tumor growth model with dynamically growing data domains given in reference [16]. This model manages to accelerate the simulation algorithm by optimizing the cellular automaton data structures while allowing data regions to be accessed by single thread at a time.

The obtained results showed better speedup in computations than the tumor growth simulation program of reference in [16]. However, there is still room for the improvement of the model. Good acceleration is obtained when using half the number of cores in the architecture. From then on a worsening on the performance is observed. In the tests we have carried out, it has been detected that the load-balancing mechanism is not fast enough to equally distribute the workload among all the threads when the tumoral mass grows. This is because the edges of the tumor are very diffuse, which causes very rapid changes in the number of cells in each region of the lattice.

Our future work will be directed towards obtaining a new implementation of our tumor growth model for GPU based on the data domain model array with dynamic growth, initially based on the sequential model in [16]. The proposed model is being improved with the inclusion of dynamic tumor growth characteristics that have been mentioned by various authors [13], such as tumor vascular prominence (angiogenesis) or tumor nutrient intake, which can be modeled using hybrid lattice-gas cellular automata [9].

References

1. Adamaztky, A., De Lacy, B., Tetsuya, A.: Reaction-Diffusion Computers. Elsevier (2005)
2. Alarcón, T., Byrne, H.M., Maini, P.K.: A cellular automaton model for tumor growth in inhomogeneous environments. J. Theor. Biol. **225**(2), 257–274 (2003)
3. Aubert, M., Badoual, M., Fereol, S., Christov, C., Grammaticos, B.: A cellular automaton model for the migration of glioma cells **3**(2), 93–100 (2006)
4. Bandman, O.: Implementation of large-scale cellular automata models on multicore computers and clusters. In: International Conference on High Performance and Simulation (HPCS), 1–5 July 2013. https://doi.org/10.1109/HPCSim.2013.6641431

5. Blecic, I., Cecchini, A., Trunfio, G.A.: Cellular automata simulation of urban dynamics through GPGPU. J. Supercomputing **65**, 614–629 (2013). https://doi.org/10.1007/s11227-013-0913-z
6. Capel-Tuñon, M.I., et al.: Towards modal modelling of biological systems. Technical report: Michigan State University, pp. 1–12 (2008)
7. Chopard, B., Droz, M.: Cellular Automata in Modeling of Physical Systems. Cambridge University Press, Cambridge (1998)
8. D'ambrosio, D., Filippone, G., Rongo, R., Spataro, W., Trunfio, G.A.: Cellular automata and GPGPU: an application to lava flow modeling. Int. J. Grid High Perform. Comput. (IJGHPC) **4**(3), 18 (2012)
9. Deutsch, A., Dorman, S.: Cellular Automata Model of Biological Patterns. Characterization, Applications and Analysis. Birkhuser (2005)
10. Enderling, H., Anderson, A., Chaplain, M., Beheshti, A., Hlatky, L., Hahnfeldt, P.: Paradoxical dependencies of tumor dormancy and progression on basic cell kinetics. Cancer Res. **69**, 8814–8821 (2009)
11. Gibson, M.J., Keedwell, E.C., Savic, D.A.: An investigation of the efficient implementation of cellular automata on multi-core CPU and GPU hardware. J. Parallel Distrib. Comput. **77**, 1125 (2015)
12. Jiao, Y., Torquato, S.: Emergent behaviors from a cellular automaton model for invasive tumor growth in heterogeneous microenvironments. PLOS Comput. Biol. **7**, Article ID: e1002314. https://doi.org/10.1371/journal.pcbi.1002314
13. Khan, M.A., Shefeeq, T., Kumar, A.: Mathematical modeling and computer simulation in cancer dynamics. Int. J. Math. Model. Simul. Appl. **4**(3), 239–254 (2011)
14. Patel, A.A., Gawlinski, E.T., Lemieux, S.K., Gatenby, R.A.: A cellular automaton model of early tumor growth and invasion: the effects of native tissue vascularity and increased anaerobic tumor metabolism. J. Theor. Biol. **213**(3), 315–331 (2001)
15. Piotrowska, M.J., Angus, S.D.: A quantitative cellular automaton model of in vitro multicellular spheroid tumour growth. J. Theor. Biol. **258**(2), 165–178 (2009)
16. Polesczuk, J., Enderling, H.: A high-performance cellular automaton model of tumor growth with dynamically growing domains. Appl. Math. **5**, 144–152 (2014)
17. Ribba, B., Alarcón, T., Marron, K., Maini, K., Agur, Z.: The use of hybrid cellular automaton models for improving cancer therapy, pp. 444–453 (2004)
18. Rybacki, S., Himmelspach, J., Uhrmacher, A.: Experiments with Single Core, Multi Core, and GPU-based computation of cellular automata. In: 2009 First International Conference on Advances in System Simulation, pp. 62–69 (2009)
19. Tomeu, A.J., Salguero, A.G., Capel, M.I.: A parallelisation tale of two languages. Ann. Multicore GPU Program. **2**(1), 81–94 (2015)
20. Trisilowati, Mallet, D.G.: Experimental modeling of cancer treatment. ISRN Oncology, **2012**, Article ID 828701 (2012). https://doi.org/10.5402/2012/828701

MOSCA: An Automated Pipeline for Integrated Metagenomics and Metatranscriptomics Data Analysis

João Carlos Sequeira, Miguel Rocha, Maria Madalena Alves,
and Andreia Ferreira Salvador(⊠)

Centre of Biological Engineering, University of Minho,
Campus de Gualtar, Braga, Portugal
asalvador@ceb.uminho.pt

Abstract. Metagenomics (MG) and Metatranscriptomics (MT) approaches open new perspectives on the interpretation of biological systems composed by complex microbial communities. Dealing with large sequencing datasets, to extract the desired information and interpret the results are big challenges associated with meta-omics studies. There are several bioinformatics pipelines for MG data analysis and less to MT. Up to date, none performs a complete analysis integrating both MG and MT data, including the assembly of reads into contigs, functional and taxonomic annotation of identified genes, differential gene expression analysis and the comparison of multiple samples. Here, we present Meta-Omics Software for Community Analysis (MOSCA) that was designed with this purpose. It integrates RNA-Seq analysis with Whole Genome Sequencing as reference. Raw sequencing reads are submitted to preprocessing for quality trimming and rRNA removal, and assembled into contigs, which afterwards are annotated by using a reference database. MOSCA performs differential gene expression and provides graphical visualization of the results and comparison of multiple samples. Validation and reproducibility of the pipeline was obtained by using simulated MG and MT datasets.

Keywords: Metagenomics · Metatranscriptomics
Bioinformatics pipeline · Community analysis · RNA-Seq
Whole genome sequencing

1 Introduction

Complex microbial communities can colonize different habitats and environments such as soil, aquatic environments, anaerobic digesters and human gut, among others. To understand the dynamics of these microbial communities within their natural environment, the use of molecular biology based technologies such as Metagenomics (MG) and Metatranscriptomics (MT) has massively increased over the last decade [1]. MG studies consist on the extraction of the

© Springer Nature Switzerland AG 2019
F. Fdez-Riverola et al. (Eds.): PACBB 2018, AISC 803, pp. 183–191, 2019.
https://doi.org/10.1007/978-3-319-98702-6_22

total DNA from a microbial community, whole genome sequencing by Next-Generation Sequencing, and finally bioinformatics data analysis and data interpretation. MG studies reveal the genetic potential of the microbial communities, giving the information about the ability of a certain community to perform a given function but do not retrieve information on the effective microbial activity. To get more insights on the activity of the microbial communities, MT studies can be performed, to inform on gene expression. Ideally, MG and MT approaches should be combined to extract more complete information. For example, it is accepted that using longer contigs, obtained by the assembly of MG/MT data, results in the identification of more annotated genes and also results in better taxonomic assignments [2].

Several tools are available to perform the analysis of Next-Generation Sequencing data, from the preprocessing of raw files to the final identification of genes. There are also several pipelines designed for the analysis of MG data, namely MOCAT2 [3] and MG-RAST [4], and few for MT analysis, like Meta-Trans [5], SAMSA2 [6] and FMAP [7]. FMAP is the only pipeline which integrates MG and MT data and performs differential analysis, but it lacks some preprocessing steps (e.g., rRNA depletion), it does not include the assembly step, nor the comparison of multiple samples. IMP [2] includes a complete preprocessing, integrates MG and MT data for the assembly step, but does not perform differential analysis. This work aims to develop and validate a pipeline, called Meta-Omics Software for Community Analysis (MOSCA), for an automated and integrated analysis of MG and MT data, performing preprocessing of raw files, assembly, annotation, differential gene expression and comparison of multiple samples. We used simulated Illumina DNA and RNA-Seq datasets, of known taxonomic and functional composition, to validate MOSCA.

2   Methods

2.1   Installation, Development and Overview of the Pipeline

MOSCA has been designed to work in a Unix system, and requires a minimum of 10 Gb of RAM, since the assemblers integrated in the workflow, MetaSPAdes [8] and Megahit [9], spend approximately 30 Gb and 10 Gb of RAM, respectively. All integrated software has to be previously installed, and their directories specified.

MOSCA[1] was developed in Python 3 and was tested in an Ubuntu server 16.04. MOSCA is fully automated to do a complete analysis of raw files generated by NGS (from MG and MT experiments), using the default parameters and tools. Nevertheless, the user can choose at the beginning between different assemblers and can also select the database used for the annotation step. MOSCA performs the complete bioinformatics analysis of MG and MT experiments. It includes the following major steps that run sequentially: preprocessing, assembly, annotation and differential gene expression between multiple samples (Fig. 1). MOSCA takes FastQ files as input, which are first submitted to quality checking by using FastQC [10]

---

[1] Available at github.com/iquasere/MOSCA.

(version 0.11.4). The majority of the preprocessing is performed with the Trimmomatic toolbox [11] (version 0.32): adapters removal with ILLUMINA-CLIP; HEADCROP and CROP to remove low quality regions at the beginning and at the end of the reads, respectively; AVGQUAL to remove reads with less quality; MINLEN to discard small reads. The arguments for CROP and HEADCROP are determined automatically based on the FASTQC report, and do not require the user input. The preprocessing of reads also includes the removal of rRNA sequences with SortMeRNA [12] (version 2.0), with reference to SILVA [13] and Rfam [14] databases. Two different assemblers are incorporated in the pipeline, MetaS-PAdes [8] (version 3.9.0) and Megahit [9] (version 1.1.1), for reconstructing the biological sequences from the trimmed reads in an iterative multi-kmer assembly. Two quality control tools for the assembly step, MetaQUAST [15] (version 4.5) and Bowtie2 [16] (version 2.2.6), report on several metrics, including N50, number of contigs for many sizes, and percentage of reads used for assembling. Gene annotation begins with the gene calling, performed with FragGeneScan [17] (version 1.15), and is followed the alignment of sequences to a FASTA database (e.g., UniProt [18]) with DIAMOND [19] (version 0.9.9). MOSCA uses the UniProt's ID mapping service application to automatically retrieve taxonomic and functional information on the annotated genes and to provide the cross-references with other databases (e.g., InterPro, CDD, RefSeq, EMBL, etc.). The output from the annotation consists in TSV files containing taxonomic and functional information and CSV files for the generation of krona plots for results visualization. MOSCA performs differential gene expression analysis by first aligning the preprocessed MT data to the MG contigs with Bowtie2, building a Gene File Format (GTF) file with a custom script, and finally building an expression matrix with htseq-count (HTSeq [20] version 0.5.4p3). DeSEQ2 [21] (version 1.18.1) is an R [22] package that organizes the expression values into graphical representations, including heatmaps denoting main genes expressed and the distance between multiple samples. Other output files are generated by MOSCA during the workflow, such as contigs, Open Reading Frame (ORF)s and alignment files produced after assembly, gene calling and annotation, and quality reports concerning preprocessing and assembly steps.

2.2   Simulation of Metagenomics and Metatranscriptomics Datasets for Validation

Illumina DNA and RNA-Seq reads were simulated to obtain datasets of known taxonomic and functional composition necessary for the validation of the pipeline. Grinder [23] (version 0.5.3) was used to simulate MG sequences (paired end Illumina reads). Grinder uses as input a FASTA file and an abundance file, which contains information on the abundance of certain genes and/or genomes in the sample. In this work, complete genome files for 22 microorganisms were downloaded from NCBI's [24] ftp site. We defined different relative abundances for the selected microorganisms. Polyester [25] (version 3.1.1) was used to simulate RNA-Seq reads. The simulation of MT reads took into consideration the abundance of the microorganisms in MG and the differential expression of some pathways, but no error models were defined for the reads. For testing and validation of MOSCA, four different experiments were generated, one corresponding to MG (MG) and three to MT collected under different conditions - a control condition (MT1), and two conditions of gene over (MT2) and under (MT3) expression. MT datasets were simulated in triplicate.

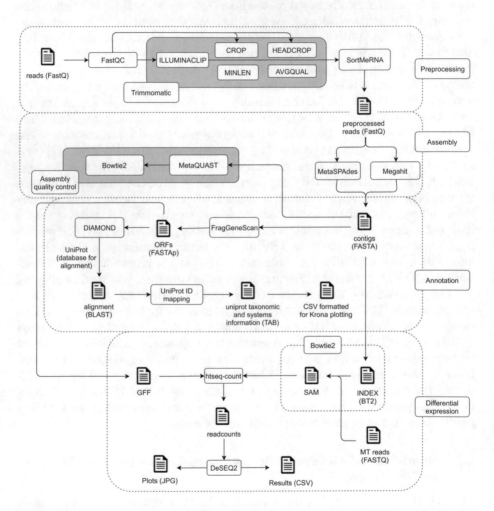

Fig. 1. Schematic overview of MOSCA. The tools integrated in MOSCA are inside the squares and input and output files are also represented.

3   Results and Discussion

Initial FastQC reports obtained for simulated FastQ files showed a decrease in quality towards the 3' end of the reads and strong biases at the 5' end towards certain nucleotides. GC content outside of normal distribution was also detected,

alongside with lower than expected average quality for reads. MOSCA's preprocessing was capable of solving the problems concerning the lower quality towards
the 3' end, and the biases in the 5' extremity for all the samples, as well as the
duplication levels in MG samples (data not shown). The results of the assembly
herein presented were obtained with MetaSPAdes. MetaQUAST reported that
2233 contigs with over 1000 bp were obtained, for an N50 of 1205, and Bowtie2
could align 66% of reads back to the contigs. From the 88199 ORFs coding to
proteins, detected with FragGeneScan, 65283 (74%) could be annotated with
DIAMOND, by using UniProt as the reference database.

The relative abundances of each taxon in simulated datasets and after analysis with MOSCA are presented in Table 1. The majority of the sequences identified in MT (76%) corresponded to the taxa selected to simulate the MT data.
However, approximately 5% of the sequences identified do not correspond to the
taxa considered in the simulation of MT nor MG. About 19% of the sequences
were identified in MT although they were not considered to simulate MT reads
(but were considered to simulate the MG reads). Approximately 30% of the
identified MG sequences were not assigned exactly to the taxa considered to

Table 1. Relative abundance (%) of the microbial taxa in simulated MG and MT
reads, and their relative abundance after analysis with MOSCA.

Taxonomic identification	Simulated for MG (%)	Identified in MG (%)	Related taxa identified in MG (%)	Simulated for MT (%)	Identified in MT (%)	Related taxa identified in MT (%)
Acinetobacter	1.586	6.644	1.061	0.000	2.585	0.268
Pseudomonas	0.783	6.293	1.070	0.000	4.384	0.270
Escherichia	1.522	5.317	2.201	0.000	8.088	2.673
Acromonas hydrophila	0.850	1.970	5.546	0.000	0.534	1.281
Geobacter	1.142	2.859	0.954	1.415	1.167	0.217
Desulfumoronadaceae	2.791	0.967	0.954	3.459	1.237	0.217
Desulfovibrio	5.708	2.095	1.390	7.074	6.394	0.291
Syntrophaceae	2.918	0.949	0.879	3.616	2.525	0.196
Syntrophobacteraceae	0.821	0.685	0.879	1.018	0.147	0.196
Bacteroides thetaiotaomicron	1.975	6.396	14.63	0.000	1.193	2.732
Clostridium botulinum	0.927	1.336	4.125	0.000	0.892	0.821
Peptococcaceae	5.074	1.994	0.758	6.288	5.853	0.158
Syntrophomonadaceae	3.806	0.318	0.758	4.717	9.536	0.158
Staphylococcus	3.928	1.726	2.641	0.000	17.00	1.033
Chloroflexus	0.888	1.515	1.584	0.000	0.283	0.290
Spirochaetia	5.074	1.047	0.170	0.000	4.557	0.026
Synergistaceae	1.776	0.418	0.170	0.000	0.707	0.026
Methanosarcina	24.97	9.115	0.642	30.94	12.74	0.131
Methanothrix	16.61	2.078	0.170	20.59	5.730	0.121
Methanomicrobiaceae	8.128	1.933	0.620	10.07	2.098	0.130
Methanospirillum	5.297	0.417	0.620	6.564	0.783	0.130
Methanobacteriales	3.425	1.847	0.088	0.000	0.000	0.121
Eukaryota	0.000	0.183	0.000	0.000	0.045	0.000
Viruses	0.000	0.000	0.000	0.000	0.024	0.000

generate the simulated MG data, but were assigned to other close taxa. The fact that these sequences are present in the simulated MG data justifies the assignment of some MT reads to the sequences that were not initially considered for the generation of the MG/MT datasets. These observations likely reflect what can be retrieved from the analysis of real MG/MT data and show that care should be taken when interpreting the results.

Figure 2 shows the relative percentage of each microorganism in the MG and MT (control), respectively, after analysis with MOSCA in krona plots. Here we can clearly visualize that the microorganisms that were defined as most active, during the generation of the simulated datasets, are the ones whose relative abundance increased. For example, *Methanothrix* and *Methanosarcina* are more abundant in MT than in MG because they were set as the most active microorganisms in the MT dataset.

Differential gene expression analysis was performed at the gene (Fig. 3A) and sample levels (Fig. 3B), resulting in two different heatmaps. In addition to the heatmaps, MOSCA retrieves tab files containing relevant information about the genes that were identified, such as their taxonomic assignment, their function, and the metabolic pathways they can be related with.

The MT triplicates clustered together when considering both the gene expression heatmap (Fig. 3A) and the heatmap showing the distance between the samples (Fig. 3B). The analysis of the simulated datasets with MOSCA clearly differentiates the datasets corresponding to the control condition (MT1 a, b, c) from the datasets corresponding to the overexpressed (MT2 a, b, c) and underexpressed condition (MT3 a, b, c). Samples MT2 and MT3 showed the biggest distance between each other and MT1 (which corresponds to the control condition) was placed between MT2 and MT3 datasets (Fig. 3B). Overexpressed genes (in red) can be found in MT2 datasets while underexpressed genes (in blue) are found in MT3 datasets. The results obtained for the differential gene expression reflect well what was defined when simulating the different datasets.

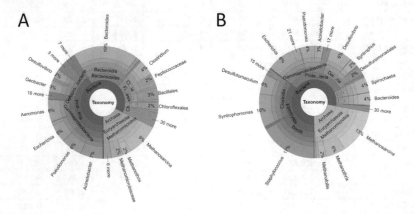

Fig. 2. Krona plots showing the relative abundance of each genus in samples MG (A) and MT control (B).

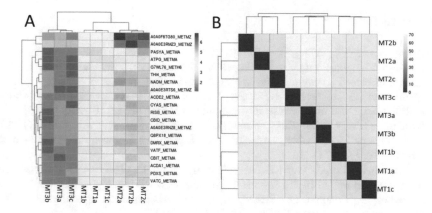

Fig. 3. Visualization of the differential gene expression results obtained with the simulated datasets. The heatmaps represent the most expressed genes (a) and the distance between samples (b) for MT1 (control condition), MT2 (overexpressed condition) and MT3 (underexpressed condition). The letters a, b and c refer to replicates.

4   Conclusions and Future Perspectives

MOSCA was developed as a command line tool for meta-omics integrated analysis, focusing on automation and independence from web servers. Nevertheless, web access is still required if UniProt's mapping service is used to obtain functional and taxonomic information. A major improvement of MOSCA is the automated trimming of reads by adjusting Trimmomatic's arguments based on FastQC's reports. Also, MOSCA performs multi-sample comparison of metatranscriptomes. MOSCA is, therefore, the third tool to integrate MG and MT analysis, after IMP and FMAP, the second to perform analysis of RNA-seq with reference to the communities' MG information, and the first to concentrate all these tasks in a single pipeline. In the future, MOSCA can be easily expanded to include metaproteomics and/or metametabolomics data analysis. Also, having a graphical interface would be useful to turn this automated pipeline into a more user friendly tool.

Acknowledgements. This study was supported by the Portuguese Foundation for Science and Technology (FCT) under the scope of the strategic funding of UID/BIO/04469/2013 unit and COMPETE 2020 (POCI-01-0145-FEDER-006684) and BioTecNorte operation (NORTE-01-0145-FEDER-000004) funded by the European Regional Development Fund under the scope of Norte2020 - Programa Operacional Regional do Norte, and by the European Research Council under the European Union's Seventh Framework Programme (FP/2007-2013)/ERC Grant Agreement no. 323009.

References

1. Zhou, J., He, Z., Yang, Y., Deng, Y., Tringe, S.G., Alvarez-cohen, L.: High-throughput metagenomic technologies for complex microbial community analysis: open and closed formats. MBio **6**(1), e02288-14 (2015)
2. Narayanasamy, S., Jarosz, Y., Muller, E.E., et al.: IMP: a pipeline for reproducible metagenomic and metatranscriptomic analyses. bioRxiv (7), 039263 (2016)
3. Kultima, J.R., Coelho, L.P., Forslund, K., et al.: Genome analysis MOCAT2: a metagenomic assembly, annotation and profiling framework. Bioinformatics **32**(16), 2520–2523 (2016)
4. Wilke, A., Bischof, J., Gerlach, W., Glass, E., et al.: The MG-RAST metagenomics database and portal in 2015. Nucleic Acids Res. **44**(D1), D590–D594 (2015)
5. Martinez, X., Pozuelo, M., Pascal, V., et al.: MetaTrans: an open-source pipeline for metatranscriptomics. Sci. Rep. **6**, 26447 (2016)
6. Westreich, S.T., Treiber, M.L., Mills, D.A., Korf, I., Lemay, D.G.: SAMSA2: a standalone metatranscriptome analysis pipeline. bioRxiv, 195826 (2017)
7. Kim, J., Kim, M.S., Koh, A.Y., et al.: FMAP: Functional Mapping and Analysis Pipeline for metagenomics and metatranscriptomics studies. BMC Bioinform. **17**(1), 420 (2016)
8. Nurk, S., Meleshko, D., Korobeynikov, A., Pevzner, P.A.: metaSPAdes: a new versatile metagenomic assembler. Genome Res. **27**(5), 824–834 (2017)
9. Li, D., Liu, C.M., Luo, R., Sadakane, K., Lam, T.W.: MEGAHIT: an ultra-fast single-node solution for large and complex metagenomics assembly via succinct de Bruijn graph. Bioinformatics **31**(10), 1674–1676 (2015)
10. Andrews, S.: FastQC: a quality control tool for high throughput sequence data (2010)
11. Bolger, A.M., Lohse, M., Usadel, B.: Trimmomatic: a flexible trimmer for Illumina sequence data. Bioinformatics **30**(15), 2114–2120 (2014)
12. Kopylova, E., Noé, L., Touzet, H.: Sortmerna: fast and accurate filtering of ribosomal RNAs in metatranscriptomic data. Bioinformatics **28**(24), 3211–3217 (2012)
13. Quast, C., Pruesse, E., Yilmaz, P., et al.: The SILVA ribosomal RNA gene database project: improved data processing and web-based tools. Nucleic Acids Res. **41**(D1), D590–D596 (2012)
14. Griffiths-Jones, S., Bateman, A., Marshall, M., Khanna, A., Eddy, S.R.: Rfam: an RNA family database. Nucleic Acids Res. **31**(1), 439–441 (2003)
15. Mikheenko, A., Saveliev, V., Gurevich, A.: MetaQUAST: evaluation of metagenome assemblies. Bioinformatics **32**(7), 1088–1090 (2015)
16. Langmead, B., Salzberg, S.L.: Fast gapped-read alignment with Bowtie 2. Nat. Methods **9**(4), 357 (2012)
17. Rho, M., Tang, H., Ye, Y.: FragGeneScan: predicting genes in short and error-prone reads. Nucleic Acids Res. **38**(20), e191 (2010)
18. UniProt Consortium: UniProt: the universal protein knowledgebase. Nucleic Acids Res. **45**(D1), D158–D169 (2016)
19. Buchfink, B., Xie, C., Huson, D.H.: Fast and sensitive protein alignment using DIAMOND. Nat. Methods **12**(1), 59–60 (2015)
20. Anders, S., Pyl, P.T., Huber, W.: HTSeqa Python framework to work with high-throughput sequencing data. Bioinformatics **31**(2), 166–169 (2015)
21. Love, M., Anders, S., Huber, W.: Differential analysis of count data – the DESeq2 package. Genome Biol. **15**, 550 (2014)

22. R Core Team: R: a language and environment for statistical computing. R Foundation for Statistical Computing, Vienna, Austria (2015)
23. Angly, F.E., Willner, D., Rohwer, F., et al.: Grinder: a versatile amplicon and shotgun sequence simulator. Nucleic Acids Res. **40**(12), 94 (2012)
24. NCBI Resource Coordinators: Database resources of the national center for biotechnology information. Nucleic Acids Res. **45**(D1), D12–D17 (2017)
25. Frazee, A.C., Jaffe, A.E., Langmead, B., Leek, J.T.: Polyester: simulating RNA-seq datasets with differential transcript expression. Bioinformatics **31**(17), 2778–2784 (2015)

Metabolite Integration Pipeline for the Improvement of Human Metabolic Models

Vítor Vieira[✉], Jorge Ferreira, Ruben Rodrigues, and Miguel Rocha

Center of Biological Engineering, University of Minho, Braga, Portugal
vvieira@ceb.uminho.pt

Abstract. Genome-scale metabolic models (GSMMs) of human cells are predictive tools with great potential for revealing important aspects of cell physiology, disease as well as for the diagnosis and treatment of diseases caused by the deregulation of metabolism. In the past decade, there have been notable efforts to reconstruct models of human metabolism, with five generic GSMMs currently available. Maintaining references to biological databases is important to allow seamless integration of models themselves and with experimental data. Still, the incorporation of external identifiers is often missed in the model reconstruction process.

In this work, we review the most relevant human GSMMs, analyze the presence of external database identifiers, extract available metabolite annotation and identifiers and create an internal database of metabolites. Using a graph-based system loaded with information from the most relevant omics data repositories, we attempt to cluster similar metabolites through database cross-referencing. With this approach, we have successfully enriched the metabolite annotation of several older GSMMs and identified common entities that could be leveraged in the future towards the creation of a unified consensus model of human metabolism.

Keywords: Genome-scale metabolic models · Human metabolism
Omics databases · Database integration

1 Introduction

Metabolism is the main mechanism responsible for the homeostasis of human cells. It can be seen as a set of biochemical reactions in a complex relationship with the genes and other factors (i.e. environment, food habits) that maintain a number of biological processes [9]. With the ongoing advances in the "omics" fields, the access to information on different layers of human cells has improved knowledge concerning their interactions and how certain disturbances can lead to disease. This type of analysis can be insightful, especially when looking at the genome-scale level [1].

Systems biology concepts such as that of Genome-scale Metabolic Models (GSMMs) are important tools to understand the link between the genome and the

© Springer Nature Switzerland AG 2019
F. Fdez-Riverola et al. (Eds.): PACBB 2018, AISC 803, pp. 192–199, 2019.
https://doi.org/10.1007/978-3-319-98702-6_23

metabolism [2,7], allowing a mathematical representation of molecular entities at a system-wide level. The first step to build a GSMM is a Genome-scale reconstruction (GENRE), which captures the set of biochemical transformations that may happen in the cell (as a metabolic network), usually curated by functional genome annotation through literature review. The main four steps for the reconstruction are the draft reconstruction, model curation, creation and validation.

Models of human metabolism have gained relevance in the last 10 years, as represented on Fig. 1. Early reconstructions focused on representing central carbon metabolism. Before the arrival of human GSMMs, Vo and colleagues presented reconstructions for the human mitochondria [14] and fibroblasts [15] that were capable of providing valuable insights on physiology and pathology.

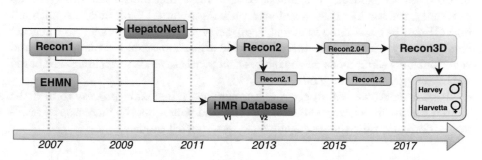

Fig. 1. Timeline of the development of human GSMMs from 2007 until the present day. This accounts only for generic models but the HepatoNet1 hepatocyte model is included as a contribution for Recon2. Each model is represented by a rounded box. Revisions and derivative models are distinguished by a black border and smaller font.

In 2007, the first two generic human GSMMs, Recon1 [4] and the Edinburgh human metabolic network (EHMN) [6], were presented. Their basis was the entire human genome, which captures the entirety of the human's metabolic capabilities. This was complemented by the creation of cell-specific models of metabolism, such as the hepatocyte model HepatoNet1 [5], which was derived from these two reconstructions and manually curated.

The Human Model Reaction (HMR) database presented by Agren et al. [7] and the Recon2 reconstruction by Thiele et al. [13] are more recent human GSMMs upon which many cell-specific studies have been developed. The authors of the latter model have since extended it, including gene-protein-reaction rules compliant with more recent versions of the human genome [8] along with a web tool to visualize and display metabolites, reactions and other pertinent information. Other authors have also built Recon2 derivatives to increase predictive capabilities on some pathways [11] or ensure better chemical balancing [10].

Recon3D is the latest development from Brunk et al. [3] (publication under review). This model is used as a basis for the Harvey and Harvetta gender-specific whole-body metabolism reconstructions recently proposed by Thiele and colleagues [12]. The authors claim this model is capable of integrating multiple organs as well as microbial communities, enabling a wide range of analyses.

Several human models provided by different sources can be integrated into a consensus metabolic model integrating biological entities (metabolites and reactions) that are shared between two or more models based on the available information in the models and external databases. However, those entities are typically identified in the models without a shared standard, which is a limitation for the integration of these models.

Metabolites present in human models can typically include information such as their name, InChI key or chemical formula, as well as identifiers for external databases such as HMDB, KEGG, PubChem, CHEBI, LipidMaps, DrugBank or BRENDA. These external databases store metabolite and reaction information as well as identifiers that can be used to connect identical components of different models. As an example, if a model A contains a metabolite M1 that with an external database identifier also found on a metabolite C1 of model B, both M1 and C1 can be considered identical, a property the can be leveraged to integrate the two models. Within the incremental identification of shared metabolites and reactions between models on the integration process, we can also infer the shared pathways present in the models.

In this work, we will apply a cross-reference integration process to find the shared metabolites across different models, and consequently, increase the available information for the metabolites present in each model.

2   Materials and Methods

For this work, we used most of the human metabolic models available. Table 1 shows an overview of the information used to perform this work, which was

Table 1. Information for the human metabolic models used in this work. The ● represents that at least one metabolite has information for that database.

	#Metabolites	ChEBI	DrugBank	HMDB	KEGG Compound	LipidMaps	PubChem Compound
Recon1	1245						
Recon1 (with drains)	1245						
HepatoNet1	712						
Edinburgh HMN	2715						
HMR (adipocyte)	6170	●			●	●	
HMR (generic)	9267						
[a]HMR2.0	3532	●		●	●		
iHuman2207	6341						
Recon2	2360	●		●	●		
Recon2 (HEK cell)	2288	●		●	●		●
Recon2.1	2360	●		●	●		
Recon2.2	2478	●	●	●	●		
Recon3D	4140	●		●	●		●
Recon3DModel	2797	●		●	●		●

[a] - The extra information for this model was obtained from the http://www.metabolicatlas.org/downloads/hmr from the .xls file.

obtained from the Systems Biology Markup Language (SBML) files of the models using the COBRApy package (with the exceptions of both Recon3D models, which were converted from Matlab format to SBML with the same package).

Figure 2(a) shows the workflow for the metabolite information for each model. Firstly, all of the SBML models are read with the COBRApy package. Since the annotation field from the metabolites was not being parsed, that information was obtained using the libSBML package. To integrate the metabolites, a matrix

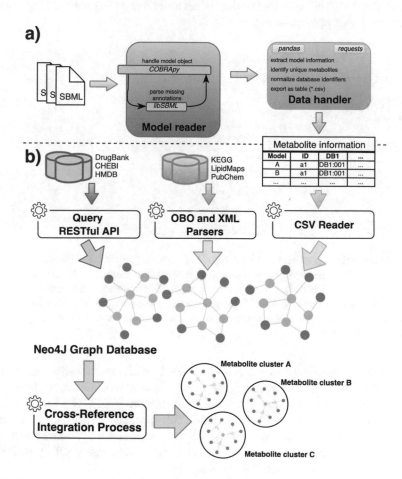

Fig. 2. Representation of the data retrieval pipeline employed in this work. (a) Pre-processing of the information of the metabolites from the SBML files; the result of this is a table containing the information for all the metabolites of the models tested. (b) Processing of information of the metabolites from external databases using Query RESTful API, OBO and XML parsers; Reading metabolite information from models CSV; Insertion of metabolite information into Neo4J graph database; Identification of metabolite clusters using a cross-reference integration process; the result of these steps is a set of metabolite clusters where each cluster groups a unique metabolite information from databases and models.

was created where each row contained a metabolite for a given model and each column contained information for the integration, i.e. identifiers for CHEBI, DrugBank, etc.

A metabolite integration tool was created using Java and a Neo4J graph database, which includes metabolite information from databases (CHEBI, Drug-Bank, HMDB, KEGG Compound, LipidMaps and PubChem Compound) and from the human metabolic models (Fig. 2b). The nodes on the graph database represent the metabolites from the databases and cross-references are represented by edges on the graph as shown in Fig. 3.

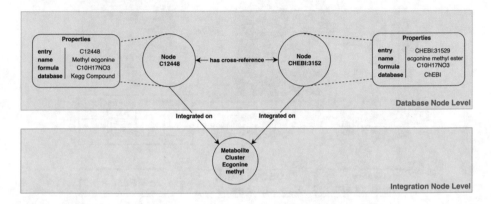

Fig. 3. Representation of the data on Neo4J database graph; The Neo4J database is separated by two main levels: database and integration node levels. Database information is loaded into metabolite nodes that contain properties with metabolite information on the database node level; The integration process connects metabolite nodes representing the same metabolite information into cluster nodes that are present in integration node level.

The graph database was initially loaded with metabolite information from metabolic models containing references to external database identifiers. These were then used to query PubChem Compound, KEGG Compound and LipidMaps databases based on their RESTful application programming interfaces (APIs) upon which the system was implemented. Finally, information from CHEBI, HMDB and DrugBank databases were loaded using OBO and XML parsers to extract information for the same identifiers.

Identification of shared metabolites on different models was performed by a cross-reference integration process. This process creates a second level of metabolite nodes on the graph database representing clusters of metabolite instances from the different databases. Each node has connections to metabolites from different models being referenced by a database identifier.

The results from this work will be available online on the following link: http://humanmodelintegration.darwin.di.uminho.pt

3 Results and Discussion

To assess the viability of the pipeline, first we show how the databases assist the integration/creation of the clusters (Figs. 4 and 5). In Fig. 4, it is possible to see that KEGG has more information than HMDB and PubChem, reflecting the larger number of the clusters with only information for that database. In terms of integration, results were good, since most compounds of each database were matched with other database; for instance, more than 80% of KEGG compounds matched HMDB and PubChem databases.

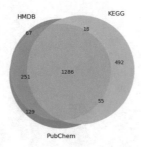

Fig. 4. Venn diagram showing the integration of the HMDB, KEGG and PubChem Compound. The numbers represent the amount of metabolite clusters containing information from the three databases, as well as clusters with overlapping references.

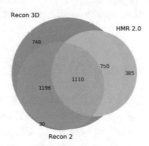

Fig. 5. Venn diagram showing the integration of the Recon3D, Recon2 and HMR2.0. The numbers represent the amount of clusters with metabolites from each model, as well as clusters that integrate two or more models.

Looking at the Fig. 5, we can clearly see that the new Recon model encompasses most of the clusters between the three models (90.1% of the clusters). Although we do not know if some metabolites from the previous Recon version are removed from this new version (since the article is only submitted), 30 clusters are not integrated with the newer one. However, we can see that the integration with HMR2.0 has improved over the Recon2. An interesting fact is that the Recon2 does not contain information about HMR2.0 and we were able to integrate at least 1860 metabolites (over 52%) from the latter model.

Figure 6 shows the evolution of the number of external identifiers to external databases for each model. Overall, the pipeline is able to significantly increase the number of metabolites which contain information to external databases. Even the most recent model had "room" for improvement, with, at least, 400 new identifiers for its metabolites.

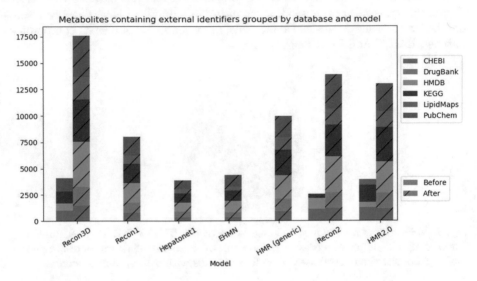

Fig. 6. Cumulative bar charts representing the number of metabolites assigned with external identifiers for each database and model before (left bar) and after (right striped bar) the integration pipeline.

4 Conclusions and Future Work

In the present work, we attempt to enrich human metabolic models with external database identifiers to improve one of their main flaws, namely their integration with other biological information repositories. With the presented pipeline, we provide a semi-automated tool to enrich the information that each model contains. Despite requiring a manual curation step, this has been shown as a good starting point to bridge these models. The next logical step for the work is the integration of the reactions of the models. With this, it could be possible to create a consensus model which would be able to incorporate all of the knowledge on the human metabolic models.

Acknowledgments. This work is co-funded by the North Portugal Regional Operational Programme, under the "Portugal 2020", through the European Regional Development Fund (ERDF), within project SISBI- RefaNORTE-01-0247-FEDER-003381. This study was also supported by the Portuguese Foundation for Science and Technology (FCT) under the scope of the strategic funding of UID/BIO/04469/2013 unit and COMPETE 2020 (POCI-01-0145-FEDER-006684) and BioTecNorte operation (NORTE-01-0145-FEDER-000004) funded by European Regional Development Fund under the scope of Norte2020 - Programa Operacional Regional do Norte. The authors also thank the PhD scholarships funded by national funds through Fundação para a Ciência e Tecnologia, with references: SFRH/BD/118657/2016 (V.V.), SFRH/BD/133248/2017 (J.F.) and SFRH/BD/131916/2017 (R.R.).

References

1. Baird, L.G., Banken, R., Eichler, H.G., et al.: Accelerated access to innovative medicines for patients in need. Clin. Pharmacol. Ther. **96**(5), 559–571 (2014)
2. Bordbar, A., Palsson, B.O.: Using the reconstructed genome-scale human metabolic network to study physiology and pathology. J. Intern. Med. **271**(2), 131–141 (2012)
3. Brunk, E., Sahoo, S., Zielinski, D.C., et al.: Recon3D enables a three-dimensional view of gene variation in human metabolism. Nat. Biotechnol. **36**(3), 272–281 (2018)
4. Duarte, N.C., Becker, S.A., Jamshidi, N., et al.: Global reconstruction of the human metabolic network based on genomic and bibliomic data. Proc. Nat. Acad. Sci. U.S.A. **104**(6), 1777–1782 (2007)
5. Gille, C., Bölling, C., Hoppe, A., et al.: HepatoNet1: a comprehensive metabolic reconstruction of the human hepatocyte for the analysis of liver physiology. Mol. Syst. Biol. **6**(411), 411 (2010)
6. Ma, H., Sorokin, A., Mazein, A., et al.: The Edinburgh human metabolic network reconstruction and its functional analysis. Mol. Syst. Biol. **3**, 135 (2007)
7. Mardinoglu, A., Nielsen, J.: Systems medicine and metabolic modelling. J. Intern. Med. **271**(2), 142–154 (2012)
8. Noronha, A.: The virtual metabolic human database: a comprehensive metabolic resource of human and human associated microbes
9. Rezzi, S., Ramadan, Z., Martin, F.P.J., et al.: Human metabolic phenotypes link directly to specific dietary preferences in healthy individuals. J. Proteome Res. **6**(11), 4469–4477 (2007)
10. Smallbone, K.: Striking a balance with Recon **2**(1), 14–17 (2013)
11. Swainston, N., Smallbone, K., Hefzi, H., et al.: Recon 2.2: from reconstruction to model of human metabolism. Metabolomics **12**, 109 (2016)
12. Thiele, I., et al.: When Metabolism Meets Physiology: Harvey and Harvetta (2018)
13. Thiele, I., Swainston, N., Fleming, R.M., et al.: A community-driven global reconstruction of human metabolism. Nat. Biotechnol. **31**(5), 419–425 (2013)
14. Vo, T.D., Greenberg, H.J., Palsson, B.O.: Reconstruction and functional characterization of the human mitochondrial metabolic network based on proteomic and biochemical data. J. Biol. Chem. **279**(38), 39532–39540 (2004)
15. Vo, T.D., Paul Lee, W.N., Palsson, B.O.: Systems analysis of energy metabolism elucidates the affected respiratory chain complex in Leigh's syndrome. Mol. Genet. Metab. **91**(1), 15–22 (2007)

A Genetic Programming Approach Applied to Feature Selection from Medical Data

José A. Castellanos-Garzón[1,2(✉)], Juan Ramos[1], Yeray Mezquita Martín[1],
Juan F. de Paz[1], and Ernesto Costa[2]

[1] IBSAL/BISITE Research Group, Edificio I+D+i USAL, University of Salamanca,
C/ Espejo s/n, 37007 Salamanca, Spain
{jantonio,juanrg,yeraymm,fcofds}@usal.es
[2] CISUC, ECOS Research Group, Pólo II - Pinhal de Marrocos,
University of Coimbra, 3030-290 Coimbra, Portugal
ernesto@dei.uc.pt

Abstract. Genetic programming represents a flexible and powerful evolutionary technique in machine learning. The use of genetic programming for rule induction has generated interesting results in classification problems. This paper proposes an evolutionary approach for logical rule induction, which is applied to clinical data. Since logical rules disclose knowledge from the analyzed data, we use such a knowledge to filter features from the target dataset. The results reached by the used dataset have been very promising when used in classification tasks and compared with other methods.

Keywords: Medical data · Feature selection · Genetic programming
Machine learning · Data mining · Evolutionary computation

1 Introduction

The exponential growth of the amount of available medical data raises the problems of efficient storage and management of information as well as disclosing useful information from the data. The problem above is a challenge in computational medicine, claiming the development of methods and tools able to transform data into medical knowledge on the underlying mechanism. Those tools (methods) allow us to go beyond a simple description of the data and provide knowledge in form of models. Through this data abstraction involving a model, we will be able to obtain predictions of systems [1–14].

There are several medical domains where machine learning techniques have been applied to discover knowledge, such as: diagnosis and prognosis, medical imaging and signal processing, planning and scheduling. Diagnosis and prognosis are the most common within this domain. Diagnosis is the process of selectively collecting information concerning a patient for its subsequent interpretation according to previous knowledge, as evidence for or against the presence or

© Springer Nature Switzerland AG 2019
F. Fdez-Riverola et al. (Eds.): PACBB 2018, AISC 803, pp. 200–207, 2019.
https://doi.org/10.1007/978-3-319-98702-6_24

absence of disorders [15]. In a prognostic process, information is also collected and interpreted through the patient. But in this case, the goal is to predict the future behavior of the patient's condition. For its predictive nature, prognosis systems are often used as tools to state medical treatments [16]. The goal of machine learning in the context of diagnosis and prognosis is knowledge discovery needed to interpret the gathered information. In some cases this knowledge has been expressed as probabilistic relationships between clinical features and the proposed diagnosis (or prognosis). In other cases a rule-based representation has been selected, so as to provide the expert with an explanation of the decision. Moreover, there are other cases where the system is designed as a black box decision maker, which is totally indifferent to the interpretation of its decisions. In summary, machine learning techniques are well suited to solve these kinds of problems thanks to their ability to search in extremely complex spaces.

In that sense, this paper proposes an approach to induce a set of logical rules across genetic programming (GP) to discover the rules representing each class of a medical dataset. Therefore, our strategy builds a rule-based classifier, which is compared with other methods and the found rules are analyzed to determine the most significant attributes from the dataset as a filtering process [17–20]. When GP is used for the rule induction problem by generating classifiers becomes one of the most important applications in this field. GP-based rule induction allows us to capture the main features of a given problem and adapts to the needs of the problem [21]. Rule-based classifiers have the property of white box, which discloses the structure of the classifier (classification rules) implying a direct process of knowledge discovery from data and the structure of their classes. Hence, GP applied to build rule-based classifiers has generated works such as [22–25]. Therefore, one of the aims of this work is to show the potential of GP applied to the rule discovery in medical datasets.

Concluding this subject, we can stress that current clinical databases store large amounts of data on patients and their medical conditions. That information stored along with the one of other patients, constitutes the scene on which machine learning is able to look for new relationships, patterns and validation of proposed hypotheses [6, 9].

2   An Approach for Rule Induction

This section describes the methodology used to build a rule-based classifier. Moreover, the genetic programming algorithm, encoding of individuals, fitness function, genetic operators are also given. Thus, we are going to start by defining logical rule as a conditional of type IF/THEN, which is represented by a set of clauses separated by conjunctions (AND) and a class k in the following way: IF $(at_1\, o_1\, val_1)$ AND $(at_2\, o_2\, val_2)$ AND $\cdots$ AND $(at_n\, o_n\, val_n)$ THEN $class = i$, where $(at_p\, o_p\, val_p)$ is called clause, at_p is an attribute of the data set, o_p a comparison operator from set $\{<, >, \leq, \geq, =, \neq\}$, val_p a value of the set of all possible values admitted by $at_p i$ whereas i is the class of the dataset covered by the rule.

The encoding used to represent individuals (in this case, rules) adopts an internal representation of a linear sequence of clauses separated by conjunctions AND. Individuals to be built by our approach follow the Michigan-style [21, 24, 26, 27]. Thus, each individual encoding a single rule (with linear chromosome) of variable length (the length of a rule is defined as its number of clauses), where each one is associated with the class it represents. The evolutionary algorithm (EA) of our approach, which is responsible for the search process for a diverse set of rules, adopts the *sequential covering* strategy for each class of a dataset [28].

2.1   Fitness Functions

This section develops the fitness functions used in the evolutionary algorithm of our approach. In this case the fitness functions defined are based on the concept of accuracy [29–31]. The accuracy of a rule is the fraction of patterns from its class, covered by the rule. Then, according to the definition above, we are going to introduce two variants of fitness functions based on accuracy. But we firstly need to define two functions which evaluate a pattern e in a rule r. Then, the first function is g acting on r and e, i.e., $g(r, e)$, which computes the number of clauses of r evaluated True when e is evaluated in r. The second function defines the evaluation of a pattern e in r ($r(e)$) in the following way:

$$r(e) = \begin{cases} 1, \text{if } r \text{ covers } e \text{ or what is the same, } e \text{ holds } r; \\ 0, \text{otherwise}. \end{cases} \tag{1}$$

Note that $g(r, e)$ evaluates the number of clauses in r holding a pattern e whereas $r(e)$ evaluates the rule to 1 (True) if it covers pattern e (all its clauses become True). Additionally, If we want to specify the class of both r and e, we write r^i and e^i respectively, where i is a class of the dataset. Finally, the two expected fitness functions are given below. For this case both fitness function define a maximization problem. The first objective of f_1 assesses accuracy based on the number of clauses turned True by patterns of the target class whereas the second objected acts as a penalty when patterns of other classes turn True clauses of the same rule. The same situation happens for f_2, but in this case, the accuracy is assessed by considering the number of patterns holding a rule r. f_1 has been created to be run in the first generations of the evolutionary algorithm where the rules have randomly been created and no pattern holds them. However, the use of f_2 makes more sense in a second stage of the evolutionary algorithm (after applying f_1) when the rendered rules have already learned.

Definition 1. *Fitness function-1.*
If D is a labeled data set with 2 classes, C_i a class of D and r^i a rule of C_i and consider $i, j \in [0, 1]$. Then we define a fitness function-1 applied to r^i as:

$$f_1(r^i) = \frac{1}{|C_i| \cdot |r^i|} \sum_{\forall e \in C_i} g(r^i, e) - \frac{1}{|D| - |C_i|} \sum_{\forall e' \in C_j, j \neq i} g(r^i, e') + 1. \tag{2}$$

Definition 2. *Fitness function-2.*
From the same conditions given in Definition 1, we define as a fitness function-2 applied to a rule r^i to:

$$f_2(r^i) = \frac{1}{|C_i|} \sum_{\forall e \in C_i} r^i(e) - \frac{1}{|D| - |C_i|} \sum_{\forall e' \in C_j, j \neq i} r^i(e') + 1. \tag{3}$$

2.2 Genetic Operators

The crossover operator used to recombine literals from two rules is similar to the classical one. It works by selecting a random position (with a uniform distribution) on two rule-parents and exchanging two segments of clauses from them to achieve two children inheriting part of the literals (genetic code) of their parents. Meanwhile, the mutation operator is responsible for providing new information to the individual generated. In this case, we provide three mutation operators:

1. *Mutation by clause*: Changes the attribute, comparison operator or value in a randomly chosen clause from the rule by others, also randomly selected;
2. *Mutation by symmetric removing*: This operator selects a position in the rule to make the removing. After that, it then selects the part of the rule to remove (on the left or right side) from the previously selected position;
3. *Mutation by addition*: Adds a new rule at the end of the current rule. The added rule is randomly created by also choosing its size in a random way.

Once the genetic operators have been defined, the evolutionary algorithm (EA) of our proposal is responsible for discovering each rule covering different parts of the data set and expecting that such rules can generalize. So the EA is executed following the general scheme given in evolutionary computation [32,33], with the particularity of introducing an elitism which is transmitted from generation to generation and tournament selection as the adopted selection method. Apart from the parameters commonly used in an EA, we added the maximum rule length.

3 Results on Clinical Data

This section describes the experiments carried out by our proposal on the clinical dataset, which has been selected from Machine Learning Repository, [34]. The dataset used deals with Statlog Heart Disease (which we call SHD)and has the following features:

- Title: Statlog Heart Disease;
- Source: Data set used for classification, Center for Machine Learning and Intelligent Systems (UCI).
  http://www.is.umk.pl/projects/datasets.html,
  http://archive.ics.uci.edu/ml/datasets/;

- Number of patterns: 270;
- Number of attributes: 13 plus the class attribute;
- Attribute type: integer and real;
- Missing attribute values: No missing values.

The class distribution is as follows:

- Absence: 150 (55.55%);
- Presence: 120 (44.45%).

3.1 Assessing Accuracy and Comparing with Other Methods

The accuracy for a rule-based classifier yielded by our approach has been computed on the SHD dataset and compared with respect to other methods on the same dataset. A stratified 10-fold cross-validation was used to measure the accuracy for all methods. The evolutionary algorithm (EA) of our approach was run in two stages. In the first stage, The EA was run by using the f_1 fitness function whereas in second stage, f_2 was used by the EA. The settings of the EA are given as follows:

```
Setting given to our proposal for the SHD dataset:
------------------------------------------------------------
Maximum length of individuals: 60
Population size: 50
Number of generations per rule (fitness function-1): 100000
Number of generations per rule (fitness function-2): 200000
Crossover probability: 0.6
Mutation probability: 0.2
```

Table 1. Names and characteristics of the methods used to classify the SHD dataset.

Method	Meaning
naive-Bayes	A classifier using neive Bayesian formula to compute the probability of each class given the values of all attributes and assuming the attribute conditional independence [35]
PCL	It is a rule-based classifier that overcomes two weaknesses of decision trees (the single coverage constraint and the fragmentation problem) by using many significant rules [36]
NeC4.5	Neural Ensemble Based C4.5. This method performs by training a neural network ensemble at first. Then, the trained ensemble is employed to generate a new training set by replacing the desired class labels of the original training patterns with those output from the trained ensemble [37]
SVM	Support Vector Machine [38]

Table 2. Comparative table of mean accuracy for five learning methods using the SHD dataset.

Method	Accuracy (%)
GP (Our approach)	86.29
naive-Bayes	84.50
PCL	83.30
NeC4.5	82.00
SVM	73.70

Then, Table 1 lists the four methods to be used in accuracy-based result comparison for the SHD dataset, whereas Table 2 lists the results of these methods compared with our approach by using mean accuracy.

4   Conclusions

This work has proposed an approach of genetic programming to induce sets of logical rules able to classify datasets. We have presented the proposal on a clinical dataset and compared with other methods. The results reached have been very promising when compared with other proposals. This proves the reliability of this approach to be used in the analysis of clinical data, which is our target data domain. Note that our proposal can be used as feature selection since the attributes appearing in the rules of the classifier are the most important and so, they discriminate the rest of attributes of the dataset.

Acknowledgments. This work has been carried out under the iCIS project (CENTRO-07-ST24-FEDER-002003), which has been co-financed by QREN, in the scope of the Mais Centro Program and European Union's FEDER.

This work has also been partially supported by the Interreg V-A Spain-Portugal Program (PocTep) and the European Regional Development Fund (ERDF) under the IOTEC project (grant 0123_IOTEC_3_E).

The research of Juan Ramos González has been co-financed by the European Social Fund and Junta de Castilla y Len (Operational Programme 2014–2020 for Castilla y Len, BOCYL EDU/602/2016).

References

1. Bandyopadhyay, S., Pal, S.K.: Classification and Learning Using Genetic Algorithms: Applications in Bioinformatics and Web Intelligence. Natural Computing Series. Springer, Heidelberg (2007). https://doi.org/10.1007/3-540-49607-6
2. Bonelli, P., Parodi, A.: An efficient classifier system and its experimental comparison with two representative learning methods on three medical domains. In: Proceedings of the 4th International Conference on Genetic Algorithms (ICGA), pp. 288–295 (1991)

3. Hong, J.H., Cho, S.B.: The classification of cancer based on DNA microarray data that uses diverse ensemble genetic programming. Artif. Intell. Med. **36**, 43–58 (2006)
4. Kumar, T.P., Iba, H.: Prediction of cancer class with majority voting genetic programming classifier using gene expression data. IEEE/ACM Trans. Comput. Biol. Bioinf. **6**, 353–367 (2009)
5. Kumar, R., Verma, R.: Classification rule discovery for diabetes patients by using genetic programming. Int. J. Soft Comput. Eng. (IJSCE) **2**, 183–185 (2012)
6. Larraaga, P., et al.: Machine learning in bioinformatics. Briefings Bioinf. **7**, 86–112 (2006)
7. Liu, K.H., Xu, C.G.: A genetic programming-based approach to the classification of multiclass microarray datasets. Bioinformatics **25**, 331–337 (2009)
8. Maulik, U., Bandyopadhyay, S., Mukhopadhyay, A.: Multiobjective Genetic Algorithms for Clustering: Applications in Data Mining and Bioinformatics. Springer, Heidelberg (2011). https://doi.org/10.1007/978-3-642-16615-0
9. Pea-Reyes, C.A., Sipper, M.: Evolutionary computation in medicine: an overview. Artif. Intell. Med. **19**, 1–23 (2000)
10. Podgorelec, V., Kokol, P., Stiglic, M.M., Hericko, M., Rozrnan, I.: Knowledge discovery with classification rules in a cardiovascular dataset. Comput. Methods Program. Biomed. **1**, 539–549 (2005)
11. Soni, J., Ansari, U., Sharma, D., Soni, S.: Intelligent and effective heart disease prediction system using weighted associative classifiers. Int. J. Comput. Sci. Eng. (IJCSE) **3**, 2385–2392 (2011)
12. Tsakonas, A., Dounias, G., Jantzen, J., Axer, H., Bjerregaard, B., von Keyserlingk, D.G.: Evolving rule-based systems in two medical domains using genetic programming. Artif. Intell. Med. **32**, 195–216 (2004)
13. Vargas, C.M.B., Chidambaram, C., Hembecker, F., Silvério, H.L.: A comparative study of machine learning and evolutionary computation approaches for protein secondary structure classification. In: Computational Biology and Applied Bioinformatics, pp. 239–258. InTech (2011)
14. Wolberg, W.H., Mangasarian, O.L.: Multisurface method of pattern separation for medical diagnosis applied to breast cytology. Proc. Natl. Acad. Sci. USA **87**, 9193–9196 (1990)
15. Lucas, P.: Analysis of notions of diagnosis. Artif. Intell. **12**(105), 295–343 (1998)
16. Lucas, P.: Prognostic methods in medicine. Artif. Intell. **15**, 105–119 (1999)
17. Ramos, J., Castellanos-Garzón, J.A., González-Briones, A., de Paz, J.F., Corchado, J.M.: An agent-based clustering approach for gene selection in gene expression microarray. Interdiscip. Sci. Comput. Life Sci. **9**, 1–13 (2017)
18. Castellanos-Garzón, J.A., Ramos, J., González-Briones, A., de Paz, J.F.: A clustering-based method for gene selection to classify tissue samples in lung cancer. In: Saberi Mohamad, M., Rocha, M., Fdez-Riverola, F., Domínguez Mayo, F., De Paz, J. (eds.) PACBB 2016. AISC, vol. 477, pp. 99–107. Springer, Cham (2016). https://doi.org/10.1007/978-3-319-40126-3_11
19. Castellanos-Garzón, J.A., Ramos, J.: A gene selection approach based on clustering for classification tasks in colon cancer. ADCAIJ Adv. Distrib. Comput. Artif. Intell. J. **4**(3), 1–10 (2015)
20. González-Briones, A., Ramos, J., De Paz, J.F.: A drug identification system for intoxicated drivers based on a systematic review. ADCAIJ Adv. Distrib. Comput. Artif. Intell. J. **4**(4), 83–101 (2015)

21. Espejo, P.G., Ventura, S., Herrera, F.: A survey on the application of genetic programming to classification. IEEE Trans. Syst. Man Cybern. Part C Appl. Rev. **40**(2), 121–144 (2010)

22. Pappa, G.L., Freitas, A.A.: Evolving rule induction algorithms with multi-objective grammar-based genetic programming. Knowl. Inf. Syst. **19**(3), 283–309 (2009)

23. Alcalá-Fdez, J., et al.: KEEL: a software tool to assess evolutionary algorithms for data mining problems. Soft. Comput. **13**, 307–318 (2009)

24. Fernández, A., García, S., Luengo, J., Bernadó-Mansilla, E., Herrera, F.: Genetics-based machine learning for rule induction: state of the art, taxonomy, and comparative study. IEEE Trans. Evol. Comput. **14**(6), 913–941 (2010)

25. Oyebode, O.K., Adeyemo, J.A.: Genetic programming: principles, applications and opportunities for hydrological modelling. World Acad. Sci. Eng. Technol. Int. J. Environ. Ecol. Geol. Min. Eng. **8**, 310–316 (2014)

26. Freitas, A.A.: A survey of evolutionary algorithms for data mining and knowledge discovery. In: Ghosh, A., Tsutsui, S. (eds.) Advances in Evolutionary Computation, pp. 819–845. Springer, Heidelberg (2002)

27. Freitas, A.A.: A review of evolutionary algorithms for data mining. In: Maimon, O., Rokach, L. (eds.) Soft Computing for Knowledge Discovery and Data Mining, Part II, pp. 79–111. Springer, Boston (2008). https://doi.org/10.1007/978-0-387-69935-6_4

28. Witten, I.H., Frank, E.: Data Mining: Practical Machine Learning Tools and Techniques, 2nd edn. Morgan Kaufmann, San Francisco (2005)

29. Flach, P.: Machine Learning: The Art and Science of Algorithms that Make Sense of Data. Cambridge University Press, Cambridge (2012)

30. Pappa, G.L., Freitas, A.A.: Automating the Design of Data Mining Algorithms: An Evolutionary Computation Approach. Springer, Heidelberg (2010). https://doi.org/10.1007/978-3-642-02541-9

31. Witten, I.H., Frank, E., Hall, M.A.: Data Mining: Practical Machine Learning, Tools and Techniques, 3rd edn. Elsevier Inc., Waltham (2011)

32. Bacardit, J., Goldberg, D.E., Butz, M.V.: Improving the performance of a pittsburgh learning classifier system using a default rule. In: Kovacs, T., Llorà, X., Takadama, K., Lanzi, P.L., Stolzmann, W., Wilson, S.W. (eds.) IWLCS 2003-2005. LNCS (LNAI), vol. 4399, pp. 291–307. Springer, Heidelberg (2007). https://doi.org/10.1007/978-3-540-71231-2_20

33. Goldberg, D.E.: Genetic Algorithms in Search, Optimization and Machine Learning. Addison-Wesley, Reading (1989)

34. Blake, C., Merz, C.: Repository of machine learning databases (UCI). Center for Machine Learning and Intelligent Systems (1998)

35. Kononenko, I., Simec, E., Robnik-Sikonja, M.: Overcoming the myopia of inductive learning algorithms with RELIEFF. Appl. Intell. **7**(1), 39–55 (1997)

36. Li, J., Wong, L.: Using rules to analyse bio-medical data: a comparison between C4.5 and PCL. In: Dong, G., Tang, C., Wang, W. (eds.) WAIM 2003. LNCS, vol. 2762, pp. 254–265. Springer, Heidelberg (2003). https://doi.org/10.1007/978-3-540-45160-0_25

37. Zhou, Z.H., Jiang, Y.: NeC4.5: neural ensemble based C4.5. IEEE Trans. Knowl. Data Eng. **16**(6), 770–773 (2004)

38. Smirnov, E., Sprinkhuizen-Kuyper, I.G., Nalbantis, I.: Unanimous voting using support vector machines. Technical report, ERIM and Universiteit Rotterdam, IKAT, Universiteit Maastricht (2004)

A DNA Sequence Corpus
for Compression Benchmark

Diogo Pratas[(✉)] and Armando J. Pinho

IEETA/DETI, University of Aveiro, Aveiro, Portugal
{pratas,ap}@ua.pt

Abstract. The progress in sequencing technologies and the increasing availability of DNA sequences from extant and extinct organisms is shaping our knowledge about species origin and development, as well as originating an improvement of the computational methods for storage and analysis purposes. Given the large volume of DNA sequences, computational models that efficiently represent diverse DNA sequences using low computational resources are very welcome. Currently, for benchmarking compression algorithms there is absence of a standard corpus that enables a wide and fair comparison. This should be a corpus that reflects the main domains and kingdoms, without being exaggerated in size and number of sequences. In this paper, we provide such DNA sequence corpus, overviewing its elements and furnishing a comparison of some of the algorithms for DNA sequence compression. The corpus is available at https://tinyurl.com/DNAcorpus.

Keywords: DNA sequences · DNA corpus · Data compression

1 Introduction

DNA sequences are sequential symbolic chains, with an alphabet of four symbols $\{A, C, G, T\}$, that can represent the genetic instructions used in the growth, development, functioning and reproduction of species. The content of DNA sequences is a past record of which evolutionary events occurred through time, being fundamental to understand the dynamics of DNA evolution and the origins of species.

However, the large volume of available DNA sequences is a serious problem for storage and analysis purposes. In storage, most of the data are discarded, while only a few part is compressed and stored, generally obtaining modest results. In analysis, the computational resources needed increase dramatically when the data is not kept organized in an efficient model. Therefore, we need better compression models, namely those which better describe the nature of the data using, as possible, the minimal computational resources.

Several algorithms have been proposed for the compression of DNA sequences, such as [1–17] (see [18] for a review). However, in the general case, their benchmarking has been addressed using specific datasets that do not reflect

© Springer Nature Switzerland AG 2019
F. Fdez-Riverola et al. (Eds.): PACBB 2018, AISC 803, pp. 208–215, 2019.
https://doi.org/10.1007/978-3-319-98702-6_25

the nature and heterogenity of the data across the diverse domains and kingdoms of species. In this line, given the high heterogenity it is easy to find or optimize a compression algorithm for a specific sequence, where others might not be so efficient.

One of the exceptions is the Manzini DNA corpus [19], available at http://people.unipmn.it/manzini/dnacorpus/dnaplain.tgz, that consists of DNA sequences belonging to four different organisms: *Saccharomyces Cerevisiae* (yeast), *Mus Musculus* (mouse), *Arabidopsis Thaliana* (plant), and *homo sapiens* (human). This DNA corpus includes mithocondrial DNA sequences (very small sequences) and very repetitive sequences (sequences from sexual chromosomes). Unfortunately, this corpus is limited to 4 species and all belonging to the eukarya domain, unreflecting the diversity of the DNA sequences.

In this paper, we make available a new DNA sequence corpus with 534,263,017 bases (509.5 MB) consisting in 15 DNA sequences, with several sizes and from different domains and kingdoms. This includes sequences from different types of viruses (phage, virus, mimivirus), archaea, bacteria and eukaryota (protozoan, fungi, amoebozoa, animalia and plant). We describe the corpus along with a consistent computational analysis. Finally, we test the DNA corpus using different compression tools.

2 The DNA Sequence Corpus

Table 1 describes the sequences present in the DNA corpus, the respective domain and/or kingdom, length and cardinality. All the sequence have been provided by the NCBI (https://www.ncbi.nlm.nih.gov/).

The DNA sequence corpus is available at https://tinyurl.com/DNAcorpus, or alternatively at http://sweet.ua.pt/pratas/datasets/DNACorpus.zip. For replicate the analysis use the scripts provided at the repository: https://github.com/pratas/DNAC.

We further describe basically the source of the species from the DNA sequence corpus as well as some analysis and relations. The first DNA sequence includes a plant species, *O. sativa*, also known as rice, and renowned for being easy to genetically modify [22]. The sequence corresponds to the assembled chromosome 1 of *O. sativa Japonica*. As it can be seen in Fig. 1 it is the third more compressible sequence from the corpus.

From the animalia kingdom, the second sequence, HoSa, corresponds to the assembled chromosome 4 of the reference human genome project. When looking in the local complexity (Fig. 1), we are able to see a major low complexity region, which is related in the centromere region of the chromosome [17]. The GaGa sequence represents the assembled chromosome 2 of *G. gallus*, commonly known as chicken, and, as it can be seen in Fig. 1, is a high complexity sequence for the average animalia [23]. The DaRe depicts the assembled chromosome 3 of *D. rerio*, also known as zebrafish [24], with the best compression ratio from the dataset. This species is particularly notable for its regenerative abilities [25]. Finally, the DrMe (*D. miranda*) represents the assembled chromosome 2 of a fruitfly, a hard sequence to compress [26].

Table 1. Description of the DNA sequences corpus. The type defines the domain and/or kingdom of the species where the sequence has been extracted. The size is in bases and '#' represents the cardinality of the alphabet.

Name	Species name	Type	Size	#
OrSa	*Oriza sativa*	Eukaryota, plant	43,262,523	4
HoSa	*Homo sapiens*	Eukaryota, animalia	189,752,667	4
GaGa	*Gallus gallus*	Eukaryota, animalia	148,532,294	4
DaRe	*Danio rerio*	Eukaryota, animalia	62,565,020	4
DrMe	*Drosophila miranda*	Eukaryota, animalia	32,181,429	4
EnIn	*Entamoeba invadens*	Eukaryota, amoebozoa	26,403,087	4
ScPo	*Schizosaccharomyces pombe*	Eukaryota, fungi	10,652,155	4
PlFa	*Plasmodium falciparum*	Eukaryota, protozoan	8,986,712	4
EsCo	*Escherichia coli*	Bacteria	4,641,652	4
HePy	*Helicobacter pylori*	Bacteria	1,667,825	4
AeCa	*Aeropyrum camini*	Archaea	1,591,049	4
HaHi	*Haloarcula hispanica*	Archaea	3,890,005	4
YeMi	*Yellowstone lake* mimivirus	Virus, mimivirus	73,689	4
BuEb	*Bundibugyo ebolavirus*	Virus	18,940	4
AgPh	*Aggregatibacter* phage S1249	Virus, phage	43,970	4

From the Entamoeba kingdom, the EnIn (*E. invadens*) is the sequence of a species that is commonly a parasite of reptiles, which is closely related with *E. histolytica*, a parasite that also infects humans. This sequence reflects high heterogenity in its local complexity (Fig. 1).

From the fungi kingdom, the ScPo (*S. pombe*) represents the complete assembled genome (3 chromosomes) of a (beer) yeast. The common information between the chromosomes is low and is, roughly, shared between the telomeres, centromeres and a few genes [27].

The last sequence from the Eukaryotic domain is the PlFa (*P. falciparum*), an unicellular protozoan parasite of humans, and the deadliest known species of *Plasmodium* that can cause malaria in humans through insects [29]. Surprisingly, in Fig. 2 we are able to see a small inter-similarity with the *Yellowstone lake* mimivirus.

From the bacteria domain we have two sequences, the EsCo (*E. coli*) and the HePy (*H. pylori*) complete genome sequences. The *E. coli* is a gram-negative bacteria that is commonly found in lower intestine of warm-blooded organisms [30] and, also, a very hard sequence to compress. The *H. pylori* is also a (small) gram-negative bacteria usually found in the stomach [31].

In the archaea domain, we have also two sequences, the AeCa (*A. camini*) and the HaHi (*H. hispanica*) complete genome sequences. The *A. camini* was isolated from a deep-sea hydrothermal vent chimney sample collected from the

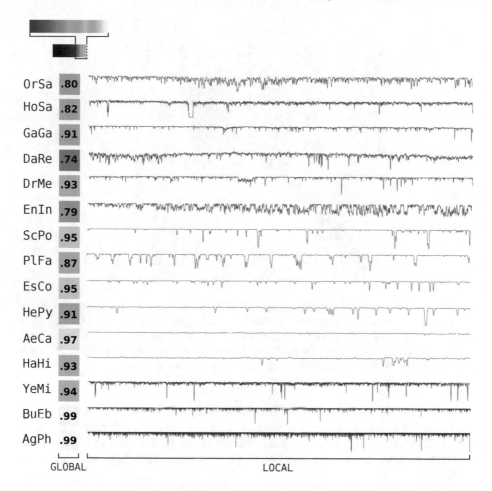

Fig. 1. Compression-based analysis of the DNA sequence corpus using the compressor GeCo [17]. The NC (Normalized Compression) [20] is given by the number of bits after compression divided by the orginal length times 2. The local complexity depicts the information content profile for each sequence [21]. Higher values stand for sequences or sub-sequences harder to compress. The complexity profiles have been filtered with window sizes cording to the sequence size. For more information and replication use the script `runAnalysis.sh`.

Suiyo Seamount, Japan, at a depth of 1385 m [32], while *H. hispanica* has been isolated in Spain. Both archaeas have unusually low restriction barriers [33].

Finally we have three viruses. The YeMi (*Yellowstone lake* mimivirus), a dsDNA algal viruses [34], the BuEb (*Bundibugyo ebolavirus*), a virus that can cause severe and often fatal hemorrhagic fever in humans and nonhuman primates [35], and AgPh (*Aggregatibacter* phage S1249), a phage that infects bacteria (*Aggregatibacter*) [36]. The three viruses are very hard to compress, specially the last two (Fig. 1).

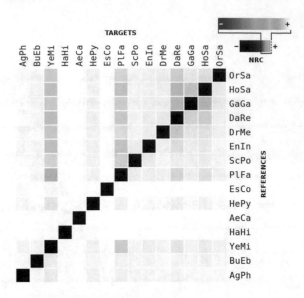

Fig. 2. Normalized Relative Compression (NRC) [28] between all the sequences of the DNA sequence corpus computed with GULL (https://github.com/pratas/GULL). The chromatic scale with colors near "+" indicate targets that are harder to be compressed relative to the respective references. The matrix is not symmetric, given the asymmetry property of the NRC. For replication use the script `runInter.sh`.

In the overall, we notice a very low inter-similarity between the sequences of the corpus (Fig. 2), showing the diversity and low common information. Moreover, in line with [20], we notice that sequences from microbiotic species, namely bacteria, archaea and viruses, are harder to compress, regardless the compressor used.

3   Compression Benchmark

For providing some results regarding the sequence DNA corpus we used two general purpose methods (Gzip and Bzip2) and, probably, the two best state-of-the-art DNA compressors (GeCo and XM) [18]. Table 2 depicts the results.

As we can see in Table 2, general purpose compressors are not able to provide efficient compression results, where XM [8] and GeCo [17] improved the compression 25% and 20% relatively to Gzip and Bzip2, respectively. This is mainly because general purpose algorithms are not prepared to deal with specific characteristics from the DNA sequences, such as heterogeneity, inverted repeats and strong stochastic variation.

In the overall, the XM algorithm was able to compress the DNA sequence corpus slightly better (below 0.4%) than GeCo, although XM used, approximately, 5 times more computational time. Regarding computational memory, XM used more RAM than GeCo.

Table 2. Number of bytes needed to represent each DNA sequence (individually) given the respective compressor (Gzip, Bzip2, GeCo, XM). We ran GeCo using "-tm 1:1:0:0/0 -tm 3:1:0:0/0 -tm 6:1:0:0/0 -tm 9:10:0:0/0 -tm 11:10:0:0/0 -tm 13:50:1:0/0 -tm 18:100:1:3/10 -c 30 -g 0.9" and XM using 50 copy experts.

Sequences	Gzip	Bzip2	GeCo	XM
OrSa	12, 247, 694	11, 576, 628	8, 671, 732	**8, 470, 212**
HoSa	53, 129, 249	48, 765, 666	**38, 877, 294**	38, 940, 458
GaGa	42, 244, 285	39, 671, 913	33, 925, 250	**33, 879, 211**
DaRe	17, 323, 427	16, 109, 731	11, 520, 064	**11, 302, 620**
DrMe	9, 214, 348	8, 683, 554	**7, 498, 808**	7, 538, 662
EnIn	7, 616, 039	7, 097, 041	5, 196, 083	**5, 150, 309**
ScPo	3, 095, 627	2, 888, 891	2, 536, 457	**2, 524, 147**
PlFa	2, 494, 880	2, 351, 428	1, 944, 036	**1, 925, 841**
EsCo	1, 341, 960	1, 252, 048	**1, 109, 823**	1, 110, 092
HePy	472, 589	442, 160	**381, 545**	384, 071
AeCa	459, 578	428, 414	**385, 640**	387, 030
HaHi	1, 115, 264	1, 030, 924	**906, 991**	913, 346
YeMi	20, 926	19, 445	17, 167	**16, 861**
BuEb	5, 762	5, 229	4, 774	**4, 642**
AgPh	12, 961	12, 066	10, 882	**10, 711**
Total	150, 794, 589	140, 335, 138	112, 986, 546	**112, 558, 213**

4   Conclusions

We have presented a new DNA sequence corpus for compression benchmark. The corpus enables a wide and fair comparison, since it reflects the main domains and kingdoms, coping with an affordable size and number of sequences.

A compression-based analysis has been made on the corpus. This included local complexity profiles and global normalized measures, such as the Normalized Compression (NC) and the Normalized Relative Compression (NRC). Finally, using the proposed corpus, we have made a small compression benchmark, showing a clear separation from GeCo and XM, regarding compression capability, to the general purpose compressors. In future works, we intent to provide a full comparison of existing open-source compressors using the proposed corpus.

Acknowledgments. This work was partially funded by FEDER (Programa Operacional Factores de Competitividade - COMPETE) and by National Funds through the FCT, in the context of the projects UID/CEC/00127/2013 & PTCD/EEI-SII/6608/2014.

References

1. Grumbach, S., Tahi, F.: Compression of DNA sequences. In: Proceedings of the Data Compression Conference, DCC-1993, Snowbird, Utah, pp. 340–350 (1993)
2. Grumbach, S., Tahi, F.: A new challenge for compression algorithms: genetic sequences. Inf. Process. Manage. **30**(6), 875–886 (1994)
3. Rivals, E., Delgrange, O., Delahaye, J.P., Dauchet, M., Delorme, M.O., Hénaut, A., Ollivier, E.: Detection of significant patterns by compression algorithms: the case of approximate tandem repeats in DNA sequences. Comput. Appl. Biosci. **13**, 131–136 (1997)
4. Chen, T., Sullivan, G.J., Puri, A.: H.263 (including H.263+) and other ITU-T video coding standards. In: Puri, A., Chen, T., (eds.) Multimedia Systems, Standards, and Networks pp. 55–85. Marcel Dekker (2000)
5. Chen, X., Li, M., Ma, B., Tromp, J.: DNACompress: fast and effective DNA sequence compression. Bioinformatics **18**(12), 1696–1698 (2002)
6. Tabus, I., Korodi, G., Rissanen, J.: DNA sequence compression using the normalized maximum likelihood model for discrete regression. In: Proceedings of the Data Compression Conference, DCC-2003, Snowbird, Utah, pp. 253–262 (2003)
7. Korodi, G., Tabus, I.: Normalized maximum likelihood model of order-1 for the compression of DNA sequences. In: Proceedings of the Data Compression Conference, DCC-2007, Snowbird, Utah, pp. 33–42, March 2007
8. Cao, M.D., Dix, T.I., Allison, L., Mears, C.: A simple statistical algorithm for biological sequence compression. In: Proceedings of the Data Compression Conference, DCC-2007, Snowbird, Utah, pp. 43–52, March 2007
9. Pinho, A.J., Ferreira, P.J.S.G., Neves, A.J.R., Bastos, C.A.C.: On the representability of complete genomes by multiple competing finite-context (Markov) models. PLoS ONE **6**(6), e21588 (2011)
10. Gupta, A., Agarwal, S.: A novel approach for compressing DNA sequences using semi-statistical compressor. Int. J. Comput. Appl. **33**(3), 245–251 (2011)
11. Zhu, Z., Zhou, J., Ji, Z., Shi, Y.: DNA sequence compression using adaptive particle swarm optimization-based memetic algorithm. IEEE Trans. Evol. Comput. **15**(5), 643–658 (2011)
12. Bose, T., Mohammed, M.H., Dutta, A., Mande, S.S.: BIND-an algorithm for lossless compression of nucleotide sequence data. J. Biosci. **37**(4), 785–789 (2012)
13. Dai, W., Xiong, H., Jiang, X., Ohno-Machado, L.: An adaptive difference distribution-based coding with hierarchical tree structure for DNA sequence compression. In: Proceedings of the Data Compression Conference, DCC-2013, pp. 371–380. IEEE (2013)
14. Li, P., Wang, S., Kim, J., Xiong, H., Ohno-Machado, L., Jiang, X.: DNA-COMPACT: DNA compression based on a pattern-aware contextual modeling technique. PLoS ONE **8**(11), e80377 (2013)
15. Guo, H., Chen, M., Liu, X., Xie, M.: Genome compression based on Hilbert space filling curve. In: Proceedings of the 3rd International Conference on Management, Education, Information and Control (MEICI 2015), Shenyang, China, pp. 29–31 (2015)
16. Xie, X., Zhou, S., Guan, J.: CoGI: towards compressing genomes as an image. IEEE/ACM Trans. Comput. Biol. Bioinf. **12**(6), 1275–1285 (2015)
17. Pratas, D., Pinho, A.J., Ferreira, P.J.S.G.: Efficient compression of genomic sequences. In: Proceedings of the Data Compression Conference, DCC-2016, Snowbird, Utah, 231–240, March 2016

18. Hosseini, M., Pratas, D., Pinho, A.J.: A survey on data compression methods for biological sequences. Information **7**(4), 56 (2016)
19. Manzini, G., Rastero, M.: A simple and fast DNA compressor. Software-Pract. Experience **34**, 1397–1411 (2004)
20. Pratas, D., Pinho, A.J.: On the approximation of the Kolmogorov complexity for DNA sequences. In: Iberian Conference on Pattern Recognition and Image Analysis, pp. 259–266. Springer (2017)
21. Pinho, A.J., Garcia, S.P., Pratas, D., Ferreira, P.J.S.G.: DNA sequences at a glance. PLoS ONE **8**(11), e79922 (2013)
22. Sales, E., Viruel, J., Domingo, C., Marqués, L.: Genome wide association analysis of cold tolerance at germination in temperate japonica rice (Oryza sativa L.) varieties. PLoS ONE **12**(8), e0183416 (2017)
23. Hudson, N., Hawken, R., Okimoto, R., Sapp, R., Reverter, A.: Data compression can discriminate broilers by selection line, detect haplotypes, and estimate genetic potential for complex phenotypes. Poult. Sci. **96**(9), 3031–3038 (2017)
24. Keck, V.A., Edgerton, D.S., Hajizadeh, S., Swift, L.L., Dupont, W.D., Lawrence, C., Boyd, K.L.: Effects of habitat complexity on pair-housed zebrafish. J. Am. Assoc. Lab. Anim. Sci. **54**(4), 378–383 (2015)
25. Goldshmit, Y., Sztal, T.E., Jusuf, P.R., Hall, T.E., Nguyen-Chi, M., Currie, P.D.: Fgf-dependent glial cell bridges facilitate spinal cord regeneration in zebrafish. J. Neurosci. **32**(22), 7477–7492 (2012)
26. Bamberger, C., Martínez-Bartolomé, S., Montgomery, M., Lavallée-Adam, M., Yates, J.R.: Increased proteomic complexity in Drosophila hybrids during development. Sci. Adv. **4**(2), eaao3424 (2018)
27. Wood, V., et al.: The genome sequence of Schizosaccharomyces pombe. Nature **415**(6874), 871–80 (2002)
28. Pinho, A.J., Pratas, D., Ferreira, P.J.S.G.: Authorship attribution using relative compression. In: Proceedings of the Data Compression Conference, DCC-2016, Snowbird, Utah, March 2016
29. Rich, S.M., Leendertz, F.H., Xu, G., LeBreton, M., Djoko, C.F., Aminake, M.N., Takang, E.E., Diffo, J.L., Pike, B.L., Rosenthal, B.M., et al.: The origin of malignant malaria. Proc. Natl. Acad. Sci. **106**(35), 14902–14907 (2009)
30. Tenaillon, O., Skurnik, D., Picard, B., Denamur, E.: The population genetics of commensal Escherichia coli. Nat. Rev. Microbiol. **8**(3), 207 (2010)
31. Eusebi, L.H., Zagari, R.M., Bazzoli, F.: Epidemiology of Helicobacter pylori infection. Helicobacter **19**(s1), 1–5 (2014)
32. Nakagawa, S., Takai, K., Horikoshi, K., Sako, Y.: Aeropyrum camini sp. nov., a strictly aerobic, hyperthermophilic archaeon from a deep-sea hydrothermal vent chimney. Int. J. Syst. Evol. Microbiol. **54**(2), 329–335 (2004)
33. Liu, H., Wu, Z., Li, M., Zhang, F., Zheng, H., Han, J., Liu, J., Zhou, J., Wang, S., Xiang, H.: Complete genome sequence of Haloarcula hispanica, a model haloarchaeon for studying genetics, metabolism, and virus-host interaction. J. Bacteriol. **193**(21), 6086–6087 (2011)
34. Zhang, W., Zhou, J., Liu, T., Yu, Y., Pan, Y., Yan, S., Wang, Y.: Four novel algal virus genomes discovered from Yellowstone Lake metagenomes. Sci. Rep. **5**, 15131 (2015)
35. Silva, R.M., Pratas, D., Castro, L., Pinho, A.J., Ferreira, P.J.S.G.: Three minimal sequences found in Ebola virus genomes and absent from human DNA. Bioinformatics **31**(15), 2421–2425 (2015)
36. Wang, J., Gao, Y., Zhao, F.: Phage-bacteria interaction network in human oral microbiome. Environ. Microbiol. **18**(7), 2143–2158 (2016)

17α-Ethinylestradiol Analysis of Endo- and Exometabolome of *Ulva lactuca* (Chlorophyta) by ¹H-NMR Spectroscopy and Bioinformatics Tools

Debora Cabral[1(✉)], Sara Cardoso[2], Silvana Rocco[3], Maurício Sforça[3],
Simone Fanan Hengeltraub[1], Cláudia Bauer[1], Miguel Rocha[2],
and Marcelo Maraschin[1]

[1] Plant Morphogenesis and Biochemistry Laboratory,
Federal University of Santa Catarina, Florianópolis, Brazil
cabraldq@gmail.com
[2] Centre of Biological Engineering, University of Minho,
Braga, Portugal
[3] Brazilian Biosciences National Laboratory (LNBio/CNPEM),
Campinas, SP, Brazil
http://ufsc.br, http://www.uminho.pt,
http://lnbio.cnpem.br

Abstract. The uptake of 17α-ethinylestradiol (EE2) by the green alga *Ulva lactuca* and its effects on the metabolome of that macroalgae have been investigated using ¹H-NMR spectroscopy. To this end, *U. lactuca in vitro* cultures have been acutely (48 h) exposed to the estrogen and both, endo- and exometabolome were investigated. Bioinformatic analyses were performed to integrate and analyze endo- and exometabolic profiles allowing the detection of EE2 into exometabolome samples and identifying very important parameters (VIPs), i.e., ¹H-NMR resonances, responsible for grouping similar *U. lactuca* metabolic profiles and biochemical phenotypes.

Keywords: *Ulva lactuca* · 17α-ethinylestradiol · NMR spectroscopy
Endometabolome · Exometabolome

1 Introduction

The concern about the frequent occurrence of 17α-ethinylestradiol (EE2) in aquatic systems and how it affects exposed organisms widely increased in the last years. Environmental metabolomics is an appealing field for assessing xenobiotics effects on natural ecosystems. Measuring changes on the composition and concentration of metabolites provide a reliable picture of cellular state of an organism and reflects the transient phenotype acquired during its interaction with the environment [1, 2]. However, assessing and analyzing the metabolic complexity of biological samples may be challenging [3]. Kuhlisch and Pohnert [4] highlighted the variety of molecules, ranging from highly polar to hydrophobic metabolites, with diverse molecular weights

© Springer Nature Switzerland AG 2019
F. Fdez-Riverola et al. (Eds.): PACBB 2018, AISC 803, pp. 216–223, 2019.
https://doi.org/10.1007/978-3-319-98702-6_26

found in marine seawater matrix. Although marine chemical ecology matured for at least three decades and generated extensive lists of marine metabolites [5], information about metabolic dynamics and composition is still elusive for some group of organisms as macroalgae [3].

Macroalgae are well referred to as efficient biomonitors for measuring bioavailable contaminants in oceans [6, 7]. Chakraborty et al. [8] pointed *U. lactuca* (Chlorophyta) as a potential biomonitor for metal pollution by determining the total metal content in the algal biomass. Henriques et al. [9] confirmed the species effectiveness in the removal of metals from seawater due to the bioconcentration of Pb, Cd, and Hg in alga tissues well above those found in the medium. Spore germination and gametophyte growth of *U. pertusa* were also proposed as sensitive monitors not only for metals (Cu, Zn, and Cd), but also for other pollutants such as formalin and TBTO (an organotin compound used as a biocide and preservative) [10]. Besides, marine photosynthetic organisms have been used in toxicological studies of exposure to hydrocarbons [11, 12], endocrine disruptors compounds (EDCs) [13–17], pesticides [18], and polycyclic aromatic hydrocarbons (PAHs), most of them with phytoplankton models.

Chaudhuri et al. [6] suggested that macroalgae would be a better indicator of heavy metals in the dissolved phase, than lipophilic compounds that would not be readily taken, due to macroalgae low lipid content. Notwithstanding, studies have evaluated macroalgae as good biological models for monitoring the toxic effects of hydrophobic molecules. For example, relevant metabolic changes resulting from *Hypnea musciformis* exposed to diesel oil [12] and *U. lactuca* to gasoline [11] have been detected, e.g., cell wall thickness and alteration of metabolic signatures such as chlorophyll or carotenoid contents, pointing to seaweed species as biological indicators of eventual impacts of petroleum derivatives on the marine environment. Schweikert et al. [18] tested the exposition of *U. pertusa* to the insecticide Coumaphos, detecting oxidative damages to the cells and increased antioxidant scavenging and GST activities.

The cosmopolitan green alga *U. lactuca* highlights by tolerating wide ranges of salinity (2–35‰), temperature, light irradiance, nutrient concentrations [19, 20] and some pollutants as described above. Its ecological plasticity and ability of rapidly respond to short-term N inputs allow it to eventually dominates shores highly impacted by anthropogenic activities [21, 22] which reinforce the convenience of its use as study model for toxicity effects of xenobiotics, bioindicators of environmental quality, and water remediation.

In this work, *U. lactuca* was exposed to EE2 for 48 h. Both, endo- and exometabolome were investigated for the first time using ^{1}H-NMR spectroscopy and bioinformatics tools. This preliminary study shed a light on unprecedented results and enabled the use of general ^{1}H-NMR resonances for discriminating treatments and understanding changes on *U. lactuca* metabolic profiles and biochemical phenotypes when exposed to xenobiotics.

2 Methods

2.1 Experimental Design

The experiment was carried using a stable *in vitro* culture of *Ulva lactuca* which has been maintained for at least 2 years under 25 ± 2 °C, 12–12 h photoperiod (280 ± 5 μmols-m^2, cool-white fluorescent lamps), and constant aeration. Experiments began two days after cutting 8 mm discs of alga thalli, which were inoculated into seawater enriched with 25% (v/v) Provasoli solution [23] as culture medium.

17α-Ethynylestradiol (EE2, $\geq 98\%$ purity, Sigma–Aldrich, Steinheim, Germany) stock solutions were prepared in ethanol 50% (v/v) and added to Provasoli medium (25%, v/v) to a final concentration 0.1 mL/L of solvent. The same concentration of ethanol was added to the control samples. Extra triplicates of control with no solvent were also included as a reference of alga basal metabolic state. Algae were exposed to 1 mg/L EE2 for 48 h and each replicate contained 5 discs (0.007 ± 0.0005 g DW) into 100 mL culture medium. Triplicates of each EE2 treatment, basal, and control discs were sampled at 0 h and 48 h, immediately washed with cold ammonium formate (0.5 M) and distilled water and freeze-dried. For exometabolome analysis, 10 mL culture medium of each replicate per treatment were assembled to form a pool and then freeze-dried.

2.2 Endo- and Exometabolome Extraction

Endometabolome. Freeze-dried biomass (20 mg) were extracted with 4.5 mL 70% methanol (v/v) containing 0.20 g/l ribitol (Sigma-Aldrich, MO, USA) as internal standard. Solution was kept from light at 4 °C for 2 h and cold-sonicated for 20 min. Extract was then centrifuged (10 min, 4000 rpm) and 1 mL supernatant of each replicate per treatment was collected and assembled to form a pool (3 mL) which was freeze-dried for further ^{1}H-NMR analysis.

Exometabolome. About 1.5 g of the resulting salt from the freeze-dried process was diluted with 30 mL of an ethyl acetate: water (1: 2, v/v) solution, stirred (10 min) and sonicated (10 min). After that, solution was poured into dark separation funnel (60 mL) and allowed to stand for 1 h, at 4 °C. The aqueous phase was discarded, and the organosolvent was vacuum dried and kept sealed (-20 °C) until ^{1}H-NMR analysis.

2.3 ^{1}H-NMR Spectroscopy

For the lyophilized extracts, 650 μl methanol-d$_4$ (Merck) containing 0.024% (w/v) sodium 3- trimethylsilyl propionate-d4 (TSP-d4) as internal standard were added, filtered (0.22 μm) and transferred to a 5 mm NMR tube for subsequent analysis. The ^{1}H-NMR spectra were recorded on an Agilent DD2 spectrometer (Brazilian Biosciences National Laboratory - Brazilian Center for Research in Energy and Materials - CNPEM), operating at 500 MHz ^{1}H frequency, at 25 °C, as described by Maraschin et al. [24]. For the endometabolome analysis, a total of 256 scans (23 min acquisition) were performed, while for the exometabolome samples, due to its lower concentration of compounds, 512 scans were collected (46 min acquisition).

^{1}H-NMR spectra pre-processing consisted of baseline correction using the Chenomx platform (v. 8.2), as the ACD software (v. 12.01) was used for internal standard reference (TSP-d4) and peak identification. Chemical shifts are expressed in δ ppm referenced to the TSP peak at $δ_{1H}$ 0.00 ppm. From the processed full spectra dataset (0.20–13.00 ppm) a resulting data matrix containing the ^{1}H-NMR chemical shifts (ppm) was generated, considering a value of 0.0002 unity as minimum normalized intensity.

2.4 Bioinformatic Analysis

Scripts using the R scientific computing system (http://www.r-project.org) were written using tools defined through the *specmine* package. Thus, hierarchical cluster analysis (HCA) was applied to the data matrix containing the ^{1}H-NMR chemical shifts (0.20–13.00 ppm) to detect eventual similarities of metabolic profiles among the investigated samples. EE2 and ribitol ^{1}H-NMR reference spectra were used for purpose of identification of those analytes in the spectra of the algal samples using the R code developed by Jacob et al. (25). Peak alignment grouped peaks together according to their position using a moving window of 0.01 ppm for matching regarding the reference compounds.

3 Results and Discussion

3.1 Identification of Xenobiotic in Complex Matrices

The effects of 17α-ethinylestradiol (EE2) on the *U. lactuca*'s metabolome were assessed on the alga tissue and culture medium extracts. As a first approach, identification of xenobiotic was aimed. By employing ^{1}H-NMR spectroscopy to the analysis of such chemically complex matrix, the used algorithm efficiently identified EE2 into exometabolome extracts using ^{1}H aromatic resonances matches (e.g. 6.3, 6.46, and 7.07 ppm) from the NMR spectra of the xenobiotic standard and algal samples. Using these approach, peaks referred to EE2 were also matched with the reference peaks of other metabolites.

EE2 was neither identified into endometabolome nor in exometabolome of control samples. The estrogen was only detected in exposed samples of exometabolome, with 72% (0 h) and 46% (48 h) matches from all identified EE2 peaks. Due to the high percentage of EE2 peaks matching with reference signals of metabolites from the complex biological samples, the ^{1}H-NMR data set was filtered considering as identified only metabolites with score higher than 0.05. In this way, resonances at 6.46 ppm only matched with EE2, while with a higher score, i.e., 0.1, other two signals at 6.52 ppm and 7.07 ppm also matched EE2 in the exometabolome of exposed samples. Alga samples collected on time 0 h presented intensities of peaks matching EE2 higher than those collected after 48 h of exposure, suggesting an eventual degradation of that compound.

Maes et al. (26) evaluated the EE2 uptake from the water-phase by the microalga *Desmodesmus subspicatus* (Chlorophyta) exposed for 72 h and estimated a half-life of the xenobiotic in cultured conditions. Authors reported EE2 transformation into more lipophilic molecules, e.g., monobrominated EE2 (Br-EE2) and dibrominated EE2 (Br2-EE2). In turn, Lai et al. [27] found EE2 stable when investigated its biotransformation by *Chlorella vulgaris* (Chlorophyta) exposed for 48 h. Degradation studies of natural and synthetic estrogens usually highlight the overall concern about longer the half-life of EE2, considering it is an estrogenic hormone with higher endocrine disruptor potency when compared with natural ones, e.g., estrone and estradiol [28]. The possibility that a cosmopolitan alga with high environmental plasticity may biotransform a potent and persistent estrogen is exciting at first. Still, if higher toxicity of subproducts as the brominated-EE2 forms [26] is confirmed, the presence of this xenobiotic in natural waters might be of bigger concern than anticipated. From our results, remains the need to confirm by which means EE2 would have been lowered in the water phase. Besides, the importance of unravelling not only the EE2 effects, but its overall interaction in the environment is highlighted.

When considering an analytical approach based on the reference spectrum of both, EE2 and Adonitol, available at an external database (i.e., Human Metabolome Database), the algorithm was not capable of identifying the targeted estrogen in the samples studied. This is because the efficiency for detecting the target compound in biologic samples is clearly lower for compounds extracted using distinct conditions from the considered samples, as well as due to different spectrum acquisition conditions (solvents and pH, e.g.).

Hierarchical clustering analysis was applied to the ^{1}H-NMR data set and two clades were observed for samples from endometabolome and from exometabolome (Fig. 1). Despite significant differences were not observed when considering exposure times or EE2 concentration, the results allow to infer a grouping tendency of exposed samples according to their endometabolome. Considering the small set of samples, non-parametric Kruskal-Wallis tests were also carried out to identify the most important resonances for endo- and exometabolome discrimination. When considering all samples, 72 out of 138 resonances showed to be relevant for grouping ($p < 0.05$). From these, p-values of 10 signals (1.24, 1.53, 7.17, 7.20, 7.55, 7.57, 7.62, 7.86, 7.94, and 7.97) were lower than 0.004, which highlights the spectral window at 7.0-8.0 ppm containing many peaks with high dissimilarities. Belhaj et al. [15] investigated metabolic changes in phytoplankton exposed to EE2 and found increments of some carbohydrates and fatty acids while photosynthetic pigments were reduced.

The capacity of the algorithm to identify EE2 in the algal samples varied according to the exposure time and concentration of EE2. The endo and exometabolome ^{1}H-NMR spectra were analyzed using HCA and in this case, the grouping of samples resulted from their endometabolome profiles over the EE2 exposure times seemed to be favoured rather than the amounts of the xenobiotic. When considering basal and control samples in the endometabolome cluster, sampling time also seemed to be more relevant than the presence of ethanol in control samples. In turn, endometabolome samples did not follow grouping for exposition times or concentrations of EE2. As for exometabolome metabolite contents were very low, eventually masking tenuous differences among the samples.

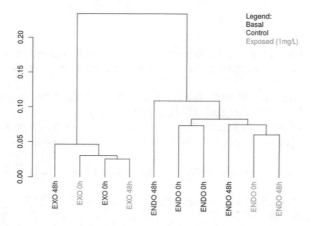

Fig. 1. Hierarchical cluster of the endo- and exometabolome samples of *Ulva lactuca*, coloured by treatments according to EE2 exposure.

4 Conclusion

The analytical approach coupled to the bioinformatics tools adopted in this study seems to be suitable for investigating EE2 in both endo- and exometabolome of *U. lactuca*. Further experimental work is required before validating a certain algorithm for identifying compounds of complex matrix using exclusively external databases as herein shown. However, the analytical approach adopted seems to be suitable for the 17α-ethinylestradiol identification in biological samples, providing that the NMR spectra are acquired at same conditions of those used by the reference analytical compound as herein noted.

Finally, our findings on the structural elucidation of the *U. lactuca*'s endo and exometabolome shed some light on certain metabolite classes EE2 may affect, addressing further metabolome studies. Whether EE2 was hydrolyzed or biotransformed by alga was not the aim of this study, but it is certainly an interesting question to be later investigated.

Acknowledgements. The authors are thankful to the Coordination for the Improvement of Higher Education Personnel (CAPES) for financial support and to the research fellowship from CNPq (grant 307099/2015-6) on behalf of M. Maraschin. This study was also supported by the Portuguese Foundation for Science and Technology (FCT) under the scope of the strategic funding of UID/BIO/04469/2013 unit and COMPETE 2020 (POCI-01-0145-FEDER-006684) and BioTecNorte operation (NORTE-01-0145-FEDER-000004) funded by European Regional Development Fund under the scope of Norte2020 - Programa Operacional Regional do Norte. This work is co-funded by the North Portugal Regional Operational Programme, under the "Portugal 2020", through the European Regional Development Fund (ERDF), within project SISBI- RefaNORTE-01-0247-FEDER-003381.

Conflict of interest statement. Authors state no conflict of interest. All authors have read the journal's Publication ethics and publication malpractice statement available at the journal's website and hereby confirm that they comply with all its parts applicable to the present work.

References

1. Lankadurai, B.P., Nagato, E.G., Simpson, M.J.: Environmental metabolomics: an emerging approach to study organism responses to environmental stressors. Environ. Rev. **21**(3), 180–205 (2013)
2. Patti, G.J., Yanes, O., Siuzdak, G.: Innovation: metabolomics: the apogee of the omics trilogy. Nat. Rev. Mol. Cell Biol. **13**(4), 263–269 (2012)
3. Gupta, V., Thakur, R.S., Reddy, C.R.K., Jha, B.: Central metabolic processes of marine macrophytic algae revealed from NMR based metabolome analysis. R. Soc. Chem. **3**, 7037–7047 (2013)
4. Kuhlisch, C., Pohnert, G.: Metabolomics in chemical ecology. Nat. Prod. Rep. **32**(7), 937–955 (2015)
5. Hay, M.E.: Marine chemical ecology: what's known and what's next? J. Exp. Mar. Biol. Ecol. **200**, 103–134 (1996)
6. Chaudhuri, A., Mitra, M., Havrilla, C., Waguespack, Y., Schwarz, J.: Heavy metal biomonitoring by seaweeds on the Delmarva Peninsula, East Coast of the USA. Bot. Mar. **50**, 151–158 (2007)
7. Torres, M.A., Barros, M.P., Campos, S.C.G., Pinto, E., Rajamani, S., Sayre, R.T., et al.: Biochemical biomarkers in algae and marine pollution: a review. Ecotoxicol. Environ. Saf. **71**(1), 1–15 (2008)
8. Chakraborty, S., Bhattacharya, T., Singh, G., Maity, J.P.: Benthic macroalgae as biological indicators of heavy metal pollution in the marine environments: a biomonitoring approach for pollution assessment. Ecotoxicol. Environ. Saf. **100**, 61–68 (2014)
9. Henriques, B., Rocha, L.S., Audia, C., Lopes, B., Figueira, P., Duarte, A.C., et al.: A macroalgae-based biotechnology for water remediation: simultaneous removal of Cd, Pb and Hg by living *Ulva lactuca*. J. Environ. Manage. **191**, 275–289 (2017)
10. Han, T., Choi, G.-W.: A novel marine algal toxicity bioassay based on sporulation inhibition in the green macroalga *Ulva pertusa* (Chlorophyta). Aquat. Toxicol. **75**, 202–212 (2005)
11. Pilatti, F.K., Ramlov, F., Schmidt, E.C., Kreusch, M., Pereira, D.T., Costa, C., et al.: *In vitro* exposure of *Ulva lactuca* Linnaeus (Chlorophyta) to gasoline – Biochemical and morphological alterations. Chemosphere **156**, 428–437 (2016)
12. Ramlov, F., Carvalho, T.J.G., Schmidt, É.C., Martins, C.D.L., Kreusch, M.G., de Oliveira, E.R., et al.: Metabolic and cellular alterations induced by diesel oil in *Hypnea musciformis* (Wulfen) J.V. Lamour. (Gigartinales, Rhodophyta). J. Appl. Phycol. 26(4), 1879–1888 (2013)
13. Shi, W., Wang, L., Rousseau, D.P.L., Lens, P.N.L.: Removal of estrone, 17alpha-ethinylestradiol, and 17beta-estradiol in algae and duckweed-based wastewater treatment systems. Environ. Sci. Pollut. Res. Int. **17**(4), 824–833 (2010)
14. Lai, K.M., Scrimshaw, M.D., Lester, J.N.: Biotransformation and bioconcentration of steroid estrogens by *Chlorella vulgaris*. Appl. Environ. Microbiol. **68**(2), 859–864 (2002)
15. Belhaj, D., Athmouni, K., Frikha, D., Kallel, M., El Feki, A., Maalej, S., et al.: Biochemical and physiological responses of halophilic nanophytoplankton (*Dunaliella salina*) from exposure to xeno-estrogen 17α-ethinylestradiol. Environ. Sci. Pollut. Res. **24**, 7392–7402 (2017)
16. Pocock, T., Falk, S.: Negative impact on growth and photosynthesis in the green alga *Chlamydomonas reinhardtii* in the presence of the estrogen 17α-ethynylestradiol. PLoS ONE **9**(10), e109289 (2014)

17. Hom-Diaz, A., Llorca, M., Rodríguez-Mozaz, S., Vicent, T., Barceló, D., Blánquez, P.: Microalgae cultivation on wastewater digestate: β-estradiol and 17α-ethynylestradiol degradation and transformation products identification. J. Environ. Manage. **155**, 106–113 (2015)
18. Schweikert, K., Burritt, D.J.: The organophosphate insecticide Coumaphos induces oxidative stress and increases antioxidant and detoxification defences in the green macroalgae *Ulva pertusa*. Aquat. Toxicol. **122–123**, 86–92 (2012)
19. Folmar, L., Hemmer, M., Hemmer, R., Bowman, C., Kroll, K., Denslow, N.: Comparative estrogenicity of estradiol, ethynyl estradiol and diethylstilbestrol in an *in vivo*, male sheepshead minnow (*Cyprinodon variegatus*), vitellogenin bioassay. Aquat. Toxicol. **49**(1–2), 77–88 (2000)
20. Messyasz, B., Rybak, A.: Abiotic factors affecting the development of *Ulva* sp. (Ulvophyceae; Chlorophyta) in freshwater ecosystems. Aquat. Ecol. **45**(1), 75–87 (2011)
21. Arévalo, R., Pinedo, S., Ballesteros, E.: Changes in the composition and structure of Mediterranean rocky-shore communities following a gradient of nutrient enrichment: descriptive study and test of proposed methods to assess water quality regarding macroalgae. Mar. Pollut. Bull. **55**(1–6), 104–113 (2007)
22. Teichberg, M., Heffner, L.R., Fox, S., Valiela, I.: Nitrate reductase and glutamine synthetase activity, internal N pools, and growth of *Ulva lactuca*: responses to long and short-term N supply. Mar. Biol. **151**, 1249–1259 (2007)
23. Starr, R.C., Zeikus, J.A.: UTEX-The culture collection of algae at the University of Texas at Austin 1993 List of Cultures. J. Phycol. **29**(s2), 1–106 (1993)
24. Maraschin, M., Somensi-Zeggio, A., Oliveira, S.K., Kuhnen, K., Tomazzoli, M., Raguzzoni, J.C., et al.: Metabolic profiling and classification of propolis samples from southern Brazil – a NMR-based platform coupled with machine learning. J. Nat. Prod. **79**, 13–23 (2016)
25. Jacob, D., Deborde, C., Moing, A.: An efficient spectra processing method for metabolite identification from ¹H-NMR metabolomics data. Anal. Bioanal. Chem. **405**, 5049–5061 (2013)
26. Maes, H.M., Maletz, S.X., Ratte, H.T., Hollender, J., Schaeffer, A.: Uptake, elimination, and biotransformation of 17α-Ethinylestradiol by the Freshwater Alga *Desmodesmus subspicatus*. Environ. Sci. Technol. **48**(20), 12354–12361 (2014)
27. Lai, K.M., Scrimshaw, M.D., Lester, J.N.: Prediction of the bioaccumulation factors and body burden of natural and synthetic estrogens in aquatic organisms in the river systems. Sci. Total Environ. **289**, 159–168 (2002)
28. Ying, G.-G., Kookana, R.S., Ru, Y.-J.: Occurrence and fate of hormone steroids in the environment. Environ. Int. **28**(6), 545–551 (2002)

Author Index

© Springer Nature Switzerland AG 2019
F. Fdez-Riverola et al. (Eds.): PACBB 2018, AISC 803, pp. 225–226, 2019.
https://doi.org/10.1007/978-3-319-98702-6

Printed in the United States
By Bookmasters